Critical Chain Project Management

For a listing of recent titles in the *Artech House Technology Management and Professional Development Library,* turn to the back of this book.

Critical Chain Project Management

Lawrence P. Leach

Artech House
Boston • London

Library of Congress Cataloging-in-Publication Data
Leach, Lawrence P.
   Critical chain project management / Lawrence P. Leach.
      p. cm. — (Artech House professional development library)
   Includes bibliographical references and index.
   ISBN 1-58053-074-5 (alk. paper)
    1. Industrial project management.   I. Title.   II. Series.
  T56.8 L34   2000
  658.4'04—dc21   99-058090
                                          CIP

British Library Cataloguing in Publication Data
Leach, Lawrence P.
   Critical chain project management. — (Artech House
   professional development library)
   1. Industrial project management
   I. Title
   658.4'04

   ISBN   1-58053-074-5

Cover design by Igor Valdman

© 2000 ARTECH HOUSE, INC.
685 Canton Street
Norwood, MA 02062

International Standard Book Number: 1-58053-074-5
Library of Congress Catalog Card Number: 99-058090

10 9 8 7

Contents

Preface

I have seen evidence from many companies attributing faster and more successful projects—and less stress on project teams—to critical chain project management (CCPM). I have also seen a number of companies, drawn by those benefits, invest in training many people and achieve little or no benefit. CCPM is already following a familiar behavior pattern of business fads. A few early adapters get great results. Others scramble to get on the bandwagon. A few succeed, but many do not. Those who do not succeed blame it on the system (a fad). The fad fades away, to be replaced by the next one.

My favorite management book is Peter Senge's *The Fifth Discipline*. Although systems thinking seems to have followed the familiar boom-and-bust cycle, organizational learning is alive and well. Buried in *The Fifth Discipline*, you will find that "to change the behavior of a system, you must identify and change the limiting factor." That is a good statement of what Eli Goldratt calls the theory of constraints (TOC).

Goldratt wrapped his theory in a love story to attract an audience wider than those who read dry management books. *The Goal*, published in 1984 by a then unknown publishing company, has sold over 2,000,000 copies in eight languages. TOC became a fad in production management and still appears to be in the early growth stage. Goldratt promises magic results from some simple thinking that he calls "uncommon sense."

Although *The Goal* was around for a dozen years by the mid-1990s, no one seemed able to relate it to project management. After I learned the TOC approach to project management, I could no longer understand how people could read *The Goal* and not see it. Goldratt tried to rectify that in 1997 with his book *Critical Chain* and is having some success.

I have come to understand that the limiting factor (which Goldratt calls the constraint) of any system involving people is their beliefs. The beliefs most important to business success are those that constitute what Harvard professor Chris Argyris calls their "theory in use." He argues that people's behavior is the only real evidence of what they believe, that their actions often conflict with their espoused beliefs.

W. Edwards Deming repeatedly stated, "In my experience, most troubles and most possibilities for improvement add up to proportions something like this: 94% belong to the system (the responsibility of management) and 6% are attributable to special causes." Special causes, also called assignable causes, are something special, not part of the system of common causes.

Argyis reports that his research found a consistent theory in use across *all* Western management, that is, a common belief system. This theory in use leads to organizational defense mechanisms that systematically prevent organizational learning. No wonder fads come and go.

It takes only a few changes in management's theory in use to make CCPM work. Management must:

1. Stop pressuring people to commit to and deliver individual project task results on specific dates (they can and should commit to project completion dates);

2. Stop causing people to multitask, enable them to work on one project task at a time until they are done, and then pass on their result as soon as it is complete with no penalty;

3. In companies with multiple projects, decide as a management team on project priorities and introduce new projects into the system only through the priority ranking;

4. Use critical chain plans for all project work;

5. Use the critical chain measurement system to make decisions on projects, including resource assignments and when to act to adjust the plan.

Unfortunately, some of that behavior requires changing the theory in use that Argyris finds extant in all organizations. If you can change those behaviors, I can promise you the benefits of CCPM. It is entirely up to your management team and is, so to speak, all in your minds.

CHAPTER 1

Begin at the beginning

Projects fail at an alarming rate. Quantitative evaluations show that as many as 30% of projects are canceled before completion, wasting all the time, money, and effort spent on them. Surviving projects usually fail to deliver the full initial project scope, or deliver late, or overrun the budget. Project delays and overruns frequently run to hundreds of percentage points. Those failures consume billions of dollars per year. They occur in all cultures and for all kinds of projects. Attempts to improve project performance create personal and organizational pain and paperwork, with little or negative impact on project performance. The field of project management has not kept pace with improvements in other areas of human endeavor, such as technology and manufacturing. This book seeks to explain why, and to put you and your organization on a path to radically improved project success.

The Project Management Institute's *Guide to the Project Management Body of Knowledge*

(PMBOK™) defines a project as "a temporary endeavor undertaken to create a unique product or service" [1]. The word *temporary* distinguishes projects from production-like endeavors. *Unique* means that projects are different from each other.

In this book, Chapters 1 through 3 make reference to the existing project system. The PMBOK™ Guide describes the system, which most project management software on the market today implements. This text considers the system described by the PMBOK™ Guide as the current theory, which uses the critical-path method (CPM) to define a project schedule. The PMBOK™ Guide alludes to other methods, but CPM is the method used most, by a wide margin. The PMBOK™ Guide describes methods to deal with uncertainty on projects through consideration of project risk. It also describes the earned value method of project measurement and control. Most large projects use project risk management and earned value, especially on projects performed for the U.S. government. Although not a specific point of guidance, most software and all the applications we have seen apply CPM using "early-start" schedules. Figure 1.1 illustrates a typical project plan using this method.

People usually distinguish projects from production operations by the quantity of the products produced and the relative amount of time on

ID	Task name
2	Plan
3	Permit
4	Site prep
5	Hole
6	Landscape
7	Drive and walks
8	
10	Foundation
11	Frame
12	Roof
13	Sheath
14	Trim
15	
17	Plumbing
18	Electrical
19	Cabinets
20	Drywall
21	Paint
22	Trim
23	Complete

Timeline columns: June (6/1, 6/8, 6/15, 6/22, 6/29), July (7/6, 7/13, 7/20, 7/27), August (8/3, 8/10, 8/17, 8/24), September (8/31, 9/7, 9/14)

Figure 1.1 A typical CPM project plan identifies the critical path and all activity start and finish dates. Most of the time, the plans use an early start schedule.

task. Projects usually produce something that is one of a kind. Production operations produce many items, all more or less similar. There is a gray area between custom-made production operations (e.g., made-to-order automobiles) and projects. I have found it interesting to observe that most people consider production operations and projects as distinctly different. A few years ago, I became interested in the system theory called the Theory of Constraints (TOC), first described by its inventor, Dr. Eliyahu Goldratt, in his book *The Goal* [2]. I recommended this book to other project managers, only to find that they could not see any relevance of the book or the theory to projects. Subsequently, I discovered a method to break the paradigm. I draw a sketch similar to Figure 1.2 and ask, "Which is this, a project or a production operation?" The reaction is interesting. Most people look puzzled at first and do not respond immediately. Then they offer, "Well, it could be either." Indeed it could. At this level, the similarity is more striking than the differences.

The actual work time in production operations is usually a very small part of the delivery time. Most people think that the actual work time (time on task) determines the overall time of project and therefore approaches 100% of the project delivery time.

1.1 Project success

Successful projects meet the needs of everyone who has an interest the project, that is, the stakeholders. All projects have a goal. Figure 1.3 illustrates that satisfying the goal normally requires satisfying three necessary conditions:

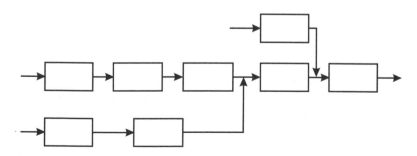

Figure 1.2 Is this a project or a production process?

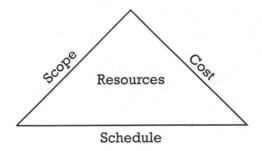

Figure 1.3 Satisfying the project goal requires three necessary conditions.

1. The *scope* sets a minimum standard for the project results.

2. The *budget* sets a maximum cost.

3. The *schedule* sets the maximum time for the project.

Figure 1.3 also shows resources, which influence all three necessary technical conditions for success.

The three necessary conditions are interdependent. The longer a project takes, the more it costs. The more a project costs, the longer it takes. The longer a project takes, the more opportunities exist to change the scope. The more the scope changes, the more cost and schedule increase. Subsequent definition of the project system explores those relationships in detail.

1.2 Defining the problem

Most scientists agree that precise definition of a problem is the most important step to a successful solution. Popper notes that "science begins with problems, and proceeds from there to competing theories which it evaluates *critically*" [3]. This text deals with the general problem of improving project success and resolving this problem: How do we design and operate a project system to satisfy customers (deliver the full scope) within the estimated (competitive) budget in the shortest time, all the time? A necessary condition to solving that problem is to motivate the people who work on the project, now and in the future.

1.2.1 How good is the current project system?

Ask yourself the following questions:

▶ Have you ever heard of projects taking longer than scheduled?

▶ Have you ever heard of a project being completed much quicker than originally scheduled, without a lot of expediting and pressure on the project team?

▶ Have you ever heard of a project going over budget?

▶ How many projects are you aware of that were completed for significantly less than the original proposed budget?

▶ Have you ever heard of projects that had to redefine their scope or specifications because they could not meet the original scope or specifications?

▶ Are customers usually delighted with project results?

In each case, the answers are usually in the undesired direction; that is, projects often underdeliver, overrun the budget, overrun the schedule, and end up with unhappy customers.

1.2.1.1 Two types of projects

Table 1.1 lists examples of two types of projects. The answers to the preceding list of questions are slightly different for the two project types. The first type is the absolute deadline–driven project. Examples include proposals and major events. Because requesters simply do not accept proposals after the specified delivery time, proposal teams rarely deliver proposals late. Management usually responds decisively to a proposal manager who spends the time and money on a proposal and delivers it late—they give the manager an opportunity to seek employment elsewhere. Likewise, although there may be much adjusting of scope and expediting, other deadline-driven projects usually happen on time. They do not delay the Olympics; they finish the stadium (somehow). People do not very often fail to have things ready for a national meeting or a prebooked trip. People rarely bow out of elections because their campaign is behind schedule. In those types of projects, the money and the scope usually change, while the schedule is held.

Table 1.1
Two Major Types of Projects

Absolute-Deadline Projects	Relative-Deadline Projects
Proposals	New-product development
Major meetings	Marketing or advertising (most)
Major events (e.g., the Olympics)	Construction
Election campaigns	Computer software
Regulatory compliance	Improvement projects
Annual budgets	Maintenance projects
Contest submissions	Research projects
Seasonal marketing	

The second type of project does not have a specific externally driven end date (although management may set one internally). All projects are performed to make money (e.g., new-product development and oil platforms) and most government projects fall into the second category, as do many improvement projects. All benefits are not lost because of project delay, just some benefits for some time. (The loss is usually understated or unknown.) In the case of projects that are not end-date driven, all three project variables (scope, schedule, and cost) may change.

1.2.1.2 Anecdotal data

Project management has a long history, which is reflected in the man-made wonders of the world. But, did they do it on schedule? Did they do it to an approved budget? Did they comply with all specifications and regulations? More and more in recent years, the answer to each of those questions is "No." Most people are aware of the major projects that have suffered from problems, for example, the new Denver Airport or the Chunnel connecting England and France. Besides being late and over budget, they also experienced scope problems. The Denver Airport baggage system did not work for a long time after the airport opened. The Chunnel had an opening ceremony but could not transport passengers. Many people are also aware of the "vaporware" problem in the software industry: Almost all software releases are later than predicted, and most have bugs in the initial release.

A recent newspaper article summarized the saga of the Denver Airport. The project was over two years late, and the cost rose from $3 billion to over $4.2 billion. The scope was not—and, as of this writing, still is not—complete. Besides the baggage problem, people cannot find their way around, leading to the spending of more than a million dollars to change the signs. Is it not likely that signs were in the original scope of the airport? The newspaper wrote the report to give the good news that the airport made a $28-million-dollar profit in 1996. Let's see: $28 million on a $4.2-billion-dollar investment works out to a return on investment (ROI) of 0.6% per year. How many investors would put their money in a project like that? Bond investors have filed a lawsuit.

Table 1.2 is found throughout the project management world and is distributed worldwide across the Internet. It is only one example of many with similar themes, attesting to the fact that projects often fail to achieve success. It is instructive to note that the effects appear to transcend all

Table 1.2
Immutable Laws of Project Management

Law 1:	No major project ever completes on time, within budget, and with the same staff that started it, and the project does not do what it is supposed to do. *Corollary 1:* The benefits will be smaller than initially estimated, if any estimates were made at all. *Corollary 2:* The system finally installed will be late and will not do what it is supposed to do. *Corollary 3:* It will cost more but will be technically successful.
Law 2:	One advantage of fuzzy project objectives is that they let you avoid embarrassment in estimating the corresponding costs.
Law 3:	The effort required to correct a project that is off course increases geometrically with time. *Corollary 1:* The longer you wait, the harder it gets. *Corollary 2:* If you wait until the project is completed, it is too late. *Corollary 3:* Do it now regardless of the embarrassment.
Law 4:	Everyone else understands the project purpose statement you wrote differently. *Corollary 1:* If you explain the purpose so clearly that no one could possibly misunderstand, someone will. *Corollary 2:* If you do something that you are sure will meet everyone's approval, someone will not like it.
Law 5:	Measurable benefits are real. Intangible benefits are not measurable, thus intangible benefits are not real. *Corollary:* Intangible benefits are real if you can prove they are real.

Table 1.2 (continued)

Law 6:	Anyone who can work effectively on a project part-time certainly does not have enough to do now. *Corollary 1:* If a boss will not give a worker a full-time job, neither should you. *Corollary 2:* If a project participant has a time conflict, the work given by the full-time boss will not suffer.
Law 7:	The greater the project's technical complexity, the less you need a technician to manage it. *Corollary 1:* Get the best manager you can. The manager will get the technicians. *Corollary 2:* The reverse of corollary 1 is almost never true.
Law 8:	A carelessly planned project will take three times longer to complete than expected. A carefully planned project will take only twice as long. *Corollary:* If nothing can possibly go wrong, it will anyway.
Law 9:	When the project is going well, something will go wrong. *Corollary 1:* When things cannot get any worse, they will. *Corollary 2:* When things appear to be going better, you have overlooked something.
Law 10:	Project teams detest weekly progress reporting because it so vividly manifests their lack of progress.
Law 11:	Projects progress rapidly until they are 90 percent complete; then they remain 90 percent complete forever.
Law 12:	If project content is allowed to change freely, the rate of change will exceed the rate of progress.
Law 13:	If the user does not believe in the system, a parallel system will be developed. Neither system will work well.
Law 14:	Benefits achieved are a function of the thoroughness of the postaudit check. *Corollary:* The prospect of an independent postaudit is a powerful incentive for a project team to deliver a good system on schedule and within budget.
Law 15:	No law is immutable.

cultures and national boundaries. Many project management books include a section on why projects fail and offer remedies to the various causes.

1.2.1.3   Quantitative data

The government is willing to compile and publish results of quantitative review of a project performance. Usually, they do not bother to publish good news on contractors, so the published information may be biased. Two quantitative examples are described next.

Following a review of major systems acquisitions (projects over $75 million) by the U.S. Department of Energy (DOE), the Government Accounting Office (GAO) reported the following [4]:

▶ From 1980 through 1996, DOE carried out 80 projects designated as major system acquisitions.

▶ Of those 80 projects, 31 were terminated prior to completion, after expenditures of over $10 billion.

▶ Only 15 of the projects were completed, most of them behind schedule and with cost overruns.

▶ Three of the 15 completed projects have yet to be used for their intended purpose.

▶ The remaining 34 projects are ongoing, many with significant cost increases and lapsed schedules.

In another report evaluating management of a recent version of a space station by the National Aeronautics and Space Administration (NASA), GAO noted the following [5]:

▶ The cost and schedule performance under the prime contract had been deteriorating for some time.

▶ Between January 1995 and April 1997, the costs associated with the schedule slippage increased from $43 million to $129 million.

▶ During that same period, the difference between the actual cost to complete specific work and the budget for that work went from a cost underrun of $27 million to a cost overrun of $291 million.

▶ As of July 1997, costs associated with the schedule slippage had increased to $135 million and the cost overrun to $355 million.

According to GAO, "The rate of decline for the cost variance is especially worrisome because it has shown no particular inclination to lessen" [5].

Your tax dollars at work! The DOE and NASA are two separate government agencies with very different projects and very different constraints, yet their performances are equally miserable.

The Department of Defense (DOD) shows similar tales of woe. Lewis [6] reports on the cancelation of the A-12 Avenger Program in 1991, which caused the loss of 9,000 jobs and entailed a lawsuit by the government for $1.35 billion in contractor overpayments. Lewis notes, "It has been acknowledged by reliable DOD sources that the C/SCSC management systems were implemented properly, and were functioning well at both the principal contractors." (C/SCSC stands for cost/schedule control system criteria, the most sophisticated project management system currently available.)

One now somewhat dated study from Australia found that construction projects completed only one-eighth of building contracts within the scheduled completion date and that the average overrun exceeded 40% [7]. Chun and Kummaraswamy reported that in a recent study of the causes of time overruns in Hong Kong construction projects [8]. The same study noted, "Delays in construction projects are still very common in most parts of the world, even with the introduction of advanced construction technologies and more effective management techniques."

Software projects are also prone to failure. Recent studies indicate that as many as 30% of software projects are canceled before they complete, and only 15% of the remaining can be considered successful in terms of the three necessary conditions.

In other words, many types of projects in many industries and in many countries (implying many cultures) seem to experience high rates of failure. The only common thread is the project system: They all use the present theory of the critical-path method, as defined by the PMBOK™ Guide. They may not all use it the same, and they may not all use it well, but they all use it.

Improving project management is, in itself, a project. To that end, several precursor conditions should be satisfied before the start of any project (in addition to the three necessary conditions of scope, budget, and schedule).

> ▶ *The right problem.* Be sure you are working on the correct problem.

> ▶ *The right solution.* Ensure that the overall objective of the project, when achieved, solves the correct problem.

> ▶ *The right design.* Develop a scope and a design that deliver an implementable solution to the correct problem.

▶ *The right implementation.* Execute the project to deliver the designed scope, achieving the objective within the planned schedule and budget.

The last point reiterates the three necessary conditions for any project.

1.2.2 The project management business

Despite the gloom-and-doom reports, many companies prosper in the business of running projects. What do these companies do that the losers are not doing? Much of the project literature would lead you to believe that they are the precious few who follow the PMBOK™ and that all you have to do to join them is do more of what you are doing and do it faster.

Successful project management companies have put in place systems that allow them to win in their environment. That environment generally includes competitors using a similar system. A competitive system does not require you to be great or even good. It does not require that your theories be right. You just have to be better than your competitors. However, the first one to put in place a dramatically improved system has the opportunity to steal the market if competitors cannot easily—or at least rapidly—match the improvement.

The current systems also must allow some of the people in the company to win, because a company needs people experienced in its system to make it work. We rarely hear about the potential impact on the rest of the people in the company or of how their suppliers get along. The model we develop of current performance predicts significant expediting, exploiting, and stress among the project participants.

One feature seems common to the project systems of successful project companies. The PMBOK™ considers it. Authors sometimes mention it in the reasons projects fail, but perhaps not often enough. Every company that succeeds in the project management business has an effective change control process. This process allows them to account for changes that happen to the project along the way and to recoup any financial impact from such changes. For example, I have worked with change control processes on major government contracts that led to thousands of formal project changes per year (too many changes to be effective, but this is just an example). An effective change process is one way to handle variation while applying the current system. Subsequent chapters reveal why it is not the best way to handle most project performance variation. An effective change control process is a necessary part of an effective project system. The critical chain method admits use of change control when necessary but dramatically reduces the number of changes.

1.2.3 Cause of the problem

Defining the problem at a high level is easy. Project managers must meet customer needs on time and at or under budget all the time. The evidence presented in this section demonstrates that the current theory does not produce this desired result. The problem is to invent a better theory that does produce the desired effect.

The Avraham Y. Goldratt Institute asks project management students, "Why is it so difficult to meet the three necessary conditions for a successful project?" The usual reasons include things like the following:

- Bad weather;

- Unforeseeable difficulties at vendors who supply equipment;

- Longer than expected time in meeting government requirements;

- An unrealistic schedule;

- Unreliable (but cheaper) vendors or contractors;

- Difficulties in matching available operators with project needs;

- Emergencies.

Such lists usually have two things in common: Whatever caused the problem is outside the control of the project manager, and the cause is some type of unexpected event.

Many project management texts include lists of the reasons projects fail. One remarkable aspect of such lists is that they list different things. Some of the lists compare the reasons for project failure viewed by different people, for example, the project manager and upper management. The lists disagree on the importance of various causes. A second remarkable aspect is that none of them suspect the project system. Two assumptions underlie many of the evaluations leading to these lists:

1. *Project work is deterministic.* The evaluations address reality as if it were possible to get accurate or precise estimates and plans. Therefore, they assume variation in the result must be caused by failure to define or operate effectively.

2. *The current project management system is effective.* This assumption leads to solutions that identify the particular part of the existing system that did not function well to cause a particular failure. None of the studies questions the effectiveness of the assumed system (which is often poorly defined in the studies themselves). None of the studies questions the assumptions underlying the assumed effective system.

One way to begin to understand project success or failure better is to look at the system, and understand some of the assumptions that underlie the current system. Following Leopold [9] (who was working in an entirely different problem domain), we can identify factors and influences that affect the success of projects. Factors are things that more or less directly affect project success in terms of scope, budget, and schedule. Success factors include:

- Selection of the right problem;
- Selection of the right solution;
- Creation of a satisfactory plan;
- An effective project control system;
- Effective project execution;
- An effective method to manage uncertainty.

Further expansion of the fourth success factor, effective project control system, leads to:

- Resource quantity;
- Resource skill;
- Resource behavior;
- The project management process;
- Project execution tools;
- Project changes.

While this list of factors may not be complete, it captures many of the items addressed in project failure studies.

In addition to the factors that seem to influence project success directly, you can also identify items that influence those factors. Project success influences internal to the project team may include:

▶ Management;

▶ Measurement;

▶ Rewards;

▶ Policies;

▶ Social norms;

▶ Variation in the processes that produce project results.

Influences external to the project team may include:

▶ Competitors;

▶ Suppliers;

▶ Client;

▶ Regulators;

▶ The physical environment;

▶ Other stakeholders (e.g., the public).

Influences may affect one or more of the factors that more directly affect project success. Table 1.3 illustrates the relationship between the influences and the factors and which influences (in my opinion) are stronger. The rows represent the factors necessary for project success. The columns represent internal and external influences on those factors. An X in a box means that the influence is of primary significance to the factor. An O in a box means that the influence has some impact on the factor. The columns with more Xs identify the most significant influences (e.g., management and policies). The columns with more blanks are less influential (e.g., competitors). Your environment may have different influences, and you may rank their significance differently. I caution you against ranking too many of the external influences as having significant

Table 1.3
Factors and Influences That Affect Project Success

Factors That Determine Project Success	Influences on Project Success Factors								
	Internal					External			
	Management	Measurement	Rewards	Policies	Social	Competitors	Suppliers	Client	Regulators
Right problem	X				X			X	O
Right solution	X				X			X	
Effective plan	X	X		X				O	O
Project control	X	X		X	X			O	O
Project execution									
Resource quantity	X			X		O	O		
Resource skill	X			X			O		
Resource behavior	X	X	X	X	X		O		
Work processes	X			X			O	O	
Tools	X			X			O		
Changes	X		X	X			O	X	
Uncertainty	X	O					X	O	X

X = significant influence
0 = some influence

influence; that could be a defense mechanism saying, "It's out of my hands." If you think it is out of your hands, you will make it so.

Note that the factors are not independent of each other. Likewise, the influences are not necessarily independent of each other. Thus, there are relationships among all the variables. The project performance system is a complex system indeed. That, combined with the sheer number of factors and influences, may explain why people attribute project failures to such a wide range of causes. For example, causes often take the form of blaming failure on the following:

▶ The customer, for not setting requirements;

▶ Senior management, for not supporting the project enough;

- ▶ Peer (resource) managers in the company, for not supporting the project enough;

- ▶ Marketing, for setting impossible requirements (including dates);

- ▶ Suppliers, for not delivering what was needed when and where it was needed;

- ▶ The system, for providing too little detail, no project plan, or ineffective change control;

- ▶ The project team, for being unmotivated, unskilled, too small, or self-serving.

System theory, which is described in Chapter 2, clarifies that influences can be more important than factors when we seek to improve a system. That is certainly true for management-controlled influences, such as the measurement and reward systems, and policies of the company. It is also true for factors that management controls directly, such as resource quantity and skill and work processes. Reasons for the relative importance of influences are that (1) the influences may affect many factors and (2) the influences may be more subject to direct intervention (change) than the factors.

The problem statement that Dr. Goldratt proposed to develop critical chain blamed poor project performance on the system. He asked, "What is it about the current system that causes so many projects to fail?" He had a good hint from his previous work with production systems and theorized that the project systems failed to manage uncertainty effectively.

1.2.4 Right solution

Over the last 40 years, many solutions have been posed to improve project management, in an attempt to better meet the customer needs on time and at or under budget. Solution trends generally are in the direction of providing more and more detail in the planning, measurement, and control of the project. Improved availability of PC-based project management systems leads to defining more tasks on projects. The software helps to automatically create a project network, define a critical path, allocate resources, and measure project performance at any level of detail. Therefore, it subtly encourages more and more detail and thus contributes to diverting attention from the important issues.

Dr. Goldratt begins *Critical Chain* with a discussion of a company wanting to reduce the time on critical development projects [10]. The company had expert consultants perform an extensive analysis; the consultants looked at the project management system and recommended many changes. In discussing the costs or time saved from all those changes, the company concluded that it would save maybe 5%, if that. Because projects fail by hundreds of percentage points, all the changes were at the wrong level.

1.2.4.1 Do more better

Earned value and derivative cost schedule control systems (CSCS, or "CS squared") [11] increase the detail of project plans and measures. The procedures companies put in place for use of systems often are many hundreds of pages long, and the number of activities in project schedules goes into the thousands. Sometimes activity duration is limited to short times, such as "no more than two weeks."

I worked with one government agency that followed the process of requiring increasingly detailed planning over a period of 20 years. Each time the agency had a project problem, it blamed some people, investigated the cause of the problem, and put in more procedures. The minimum time to do a project crept up to almost seven years, not including the time to do the project! Thus, the agency built in seven years of planning time before the start of any project. There are engineering studies and conceptual design reports and independent cost estimates and validated cost schedule control systems. Yet the costs and the schedule demands of projects continued to rise, and more and more projects failed to meet technical requirements. In one case, the agency canceled a project after having spent over a billion dollars on it. Other projects are tens of years late.

One study showed it cost the agency four times as much per square foot as local construction by nongovernment purchasers to build a simple office building. Projects were having larger and larger crises, in which they would "rebaseline," yielding new cost and schedule estimates several times (usually three or more times) the original estimates. They canceled larger and larger projects because the need was gone before the project was over or because the newly projected cost and schedule changed the cost-benefit equation to where the project no longer made sense. That is the problem the agency was trying to solve in the first place.

Is the world changing that much? On the other hand, could it be that our solutions are actually making things worse, not better?

Let's review the logic of the "do more better" approach. If your objective is to reliably complete projects to the scope, schedule, and cost, you must define those requirements accurately. To define requirements accurately, you must add detail to your project plans, because previous projects failed to deliver at the current level of detail. That logic seems to make sense and to be in line with literature that attributes project failure to inadequate requirements or insufficient detail in the project plan.

The "do more better" approach frequently leads to project plans with thousands of activities. We recently worked with clients who were rather proud of the fact that their project plan contained more than 15,000 activities. Consider a much more modest project plan that contains a mere 100 activities. The average size of an activity in that plan (measured in dollars, person-days, or even task-days) would be, by simple math, 1% of the total project (by comparable measure). Most project managers would be happy to have their project come in within 1% of plan. The problem with project success must involve something that causes variations of far more than 1%. Therefore, it is evident that increasing plan detail beyond 100 activities is not going to improve project success.

Sometimes people defend the more detailed method by suggesting that the problem, even though much bigger than 1%, is that they miss something in their plan. You are not likely to find the missing 20% inside the 1% chunks of the project. Looking inside the 1% for the big hitters reminds me of the story about the drunk who lost his car keys in the alleyway but is looking for them under the streetlight because, "I can't see anything over there in the dark." If you are worried about missing big chunks, you would do far better to examine the spaces between the 100 activities you have rather than break the defined activities into greater and greater detail.

Some of the literature that poses causes and solutions to project problems also offers anecdotal evidence that solutions improved project success in one or more subsequent projects. While such evidence is interesting, it does little to prove that the solution has really found the cause of failure in project systems, for the following reasons:

▶ *Theory of knowledge.* One or more successful cases do not prove a theory (discussed in Chapter 2).

▶ *The environment.* If the system had poor practices to begin with, any degree of discipline is likely to cause an improvement.

▶ *Regression to the mean.* A particularly bad performance is likely to be followed by a better performance.

▶ *The Hawthorne effect.* In this psychological effect, workers singled out to try new methods respond positively to any change, including changing back to conditions that existed before the experiment.

In other words, the posed theories have not been subject to effective experimental tests.

1.2.4.2 Uncertainty

Everyone knows that project tasks have a certain amount of inherent uncertainty. The very definition of a project says you have not done this task before, or at the least, you have not performed all the tasks the same way you will in this particular project. To complete the project successfully, you must account for such uncertainty. People's ability to estimate off the cuff varies depending on a number of factors. There is substantial evidence to indicate that people tend toward overconfidence in their belief in the accuracy of their estimates [12]. It is unlikely that most project tasks can be estimated better than ±20%.

As part of our training classes, we have people estimate a simple task: going to a local store and buying a specified object. If necessary, we tell them where the store is. Nearly all the participants in the exercise agree that the task is much simpler than most of their project tasks. They also agree that the ability of the other people in the room to estimate the task should be as good as or better than the ability of their project estimators to calculate project estimates. The range of the estimates usually is several hundred percentage points of the mean, and the standard deviation is usually on the order of 30% of the mean. Figure 1.4 illustrates typical results from this exercise.

Figure 1.5 illustrates the expected general behavior of the accuracy of a single task estimate as a function of the amount of effort put into creating the estimate. The accuracy scale presents the accuracy as a percentage of the mean estimate, so a perfectly accurate estimate has an accuracy of zero. An estimate with no effort at all should have an accuracy of at least 100% on the down side and could be orders of magnitude (hundreds of

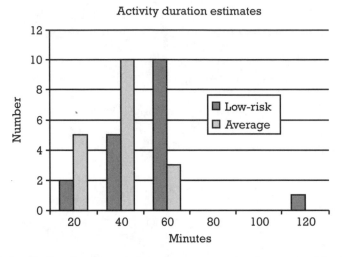

Figure 1.4 Estimate uncertainty for a very simple project task illustrates the typical range of real uncertainty.

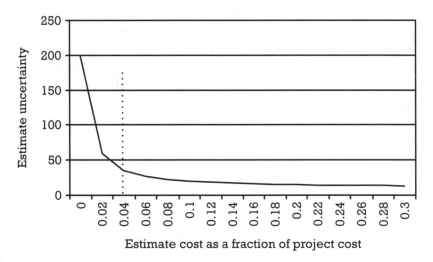

Figure 1.5 Estimate accuracy generally increases with the effort applied to the estimate, up to a limit determined by the process involving the subject of the estimate.

percentage points) too low. The curve illustrates that the accuracy should generally improve as more effort is put toward the estimate. A lower limit usually limits improvement due to the inherent variation in the process that will produce the task result. That lower limit, described further in Chapter 2, is called common cause variation. No matter how much more effort you put into the estimate, you can never do better than the common cause variation of the process that produces the result of the task. You can reduce common cause variation only by changing the task process.

Consider the two regions in Figure 1.5 divided by the vertical dotted line. To the right of the line, adding more effort to the estimate does not significantly improve the accuracy of the estimate. To the right of the line, reducing the effort does not have much impact on uncertainty. Estimates to the left of the line show increasing sensitivity to the amount of effort applied. Small reductions in the applied effort greatly increase the uncertainty of the estimate, and small increases in the effort significantly improve the estimate.

The effect on overall plan uncertainty that will obtain from adding more tasks to your project plan depends on the region in which you are operating. Assuming a fixed level of investment in the estimate, if you are well to the right of the line, adding more tasks (which reduces the effort per task) may increase the accuracy of the overall plan. The reason is that the accuracy of the overall plan improves as the plan is divided into more equal-sized pieces, if the accuracy of the individual pieces is the same.

If the amount of estimating effort you can afford puts you near or to the left of the vertical line in Figure 1.5, adding more tasks to your plan can decrease the accuracy of the overall plan. The reason is that the increasing uncertainty in each task estimate can be much greater than the statistical benefit of more individual tasks.

Adding more tasks to a project plan increases the number of potential task connections much faster than the number of tasks you add. For example, if you add one task to a plan with 100 tasks, you only add one task. However, you add the potential for 200 additional connections because each task in the existing plan is a potential predecessor and a potential successor to the task you just added. The additional potential relationships greatly increase the probability of errors in the project task network as the number of tasks in the plan increases.

A cause that might be deduced from project failure due to the alleged causes of project failure posed so far is that uncertainty causes projects to fail. If that were the case, then all projects with uncertainty should fail. Based on the definition of a project and our understanding of the real world, all projects have uncertainty, and, therefore, all projects should fail. However, not all projects fail. Furthermore, there is evidence that some projects succeed despite extreme uncertainty. In *Critical Chain*, Goldratt describes an airplane project that defied that prediction: The designers developed an airplane with unprecedented capabilities in eight months, instead of the 10 years such developments normally take. There are other cases. The United States succeeded in meeting President Kennedy's objective to put a man on the moon by the end of the 1960s, one of the most uncertain projects ever undertaken. Creation of the atomic bomb was a similarly uncertain project completed in a remarkably short time.

A cornerstone of the scientific method is that scientists can never prove that any scientific theory or law will continue to work in the future; they can, however, disprove a theory with just one proper test. More than one instance proves that uncertainty itself cannot be the cause of project failure.

If simple uncertainty does not meet the test of explaining project failure, can the theory be modified to fit the known evidence? Some projects use different ways to manage uncertainty. For example, the Apollo project managed risk by hiring three companies to produce three different solutions for high-risk developments. One solution was chosen as the primary path, and the other two were backups in case the primary path failed. NASA planned on much test and retest (and they had plenty of spectacular failures along the way). While that is an expensive way to manage uncertainty, it worked. Goldratt used thinking like that to pose the hypothesis that it is failure to effectively manage uncertainty that causes most projects to fail. Chapter 3 examines that hypothesis in depth. If Goldratt is right, the direction of the solution is to create a different project system more able to manage uncertainty.

1.2.5   Right execution

Right execution refers to execution of the solution to the problem. Improvement to the project system is itself a project.

In his pamphlet "My Saga to Improve Production," Dr. Goldratt noted the following:

> It took me some time to figure it out, but at last I couldn't escape
> the simple explanation: the efforts to install the software distracted the

plant people from concentrating on the required changes—the changes in fundamental concepts, measurements and procedures. [13]

A similar phenomenon occurs in many efforts to improve perform-ance of the project system. The usual solutions are along the line of doing the current system better, which many people interpret as more detail and more documentation. That often involves installing new project or database software. Such solutions distract people further from perform-ing the project and seldom seem to improve much. Of course, better implementation of a flawed system is unlikely to improve much anyway. Chapter 10 provides an effective plan to implement the critical chain project system.

1.3 Success with critical chain

Now that we have defined the problem and substantiated the claim that the current theory is in need of improvement, the next step requires cre-ating a new theory (of the project system): critical chain project manage-ment (CCPM). Expectations for the theory are that it will, subject to critical evaluation, consistently achieve project success. It should explain both past success and failure and provide testable predictions of future performance. Preliminary experience with the new theory shows benefits that exceed the minimal performance requirements for the new theory but that the theory can explain. Those benefits (compared to the present critical path theory) are the following.

- Improved project success:
  - Projects completed on time all the time;
  - Projects delivered full scope;
  - Project cost under budget;
  - Improved market position and business growth.
- Reduced project duration:
  - Projects completed in half the time (or less) of previous similar projects;
  - Individual project plans reduced by at least 25%;

- Multiple project durations reduced by larger amounts;
- Project changes reduced;
- Early returns for commercial projects;
- Reduced payback periods for investment projects.

▶ Increased project team satisfaction:

- Reduced confusion from multitasking;
- Ability to focus on one task at a time;
- Reduced changes;
- Reduced rework;
- Reduced pressure from multiple project managers;
- Win-lose task completion (date-driven task pressure) eliminated;
- Buffer reporting used by individuals to decide task priority;
- Reduced insertion of new priority tasks.

▶ Simplified project measurement:

- Quick and easy plan status;
- Real-time project status; no need to wait for financial reports;
- Immediate focus by buffer, chain, and task provided by status;
- Decisions defined by buffer report;
- Focus of buffer reporting on management priority decisions (reflected in the buffers by staggering project start).

▶ Simplified project management:

- Clear focus for project manager (critical chain, reduced early start);
- Simplified project plans reduce paperwork;
- Simplified project status reporting;
- Whether to plan or act decided by measurement;
- Resource priorities decided by measurement.

> ◗ Increased project throughput with same resource:

>> ◗ Reduced resource demand conflicts;

>> ◗ More projects completed faster for the same level of resources;

>> ◗ Less need to hire new critical resources;

>> ◗ Less delay due to resources;

>> ◗ Improved project cash flow;

>> ◗ Improved ROI.

Evidence of other users often gives people confidence to try new ideas. The present CPM project paradigm has been in force for over 40 years, making change hard for many people to accept. More and more companies, small and large, are demonstrating success with CCPM. Several examples illustrate that success. (As will be discussed later, these success examples do not "prove" the new theory; they only provide confidence that it is not fatally flawed.)

> ◗ *Honeywell Defense Avionic Systems (DAS) is experimenting with critical chain. A recent internal article noted the following for a project they named RNLAF.* "The RNLAF team was asked by the customer to deliver something we originally scheduled to take 13 months to deliver—and the team did it in six months ... The team is experimenting with a new way of scheduling the program using critical chain concepts. Boeing has read the book, and is supporting the concept." [13]

> ◗ *Lucent Technologies.* Lucent Technologies has adopted CCPM as their primary tool for project management. (The author provides Lucent training and implementation assistance.) "In 1996, Lucent Technologies Advanced Technology Systems, now part of General Dynamics, was told by a sister organization that the yearlong project being considered was an impossibility ... The project was used as a pilot effort, to evaluate TOC project management. The project was completed in June, 1997, with buffer to spare." [14]

> ◗ *Harris.* Harris recently decided to use CCPM to build a new eight-inch semiconductor, wafer plant. The largest previous wafer was 6 inches in diameter. Total investment for a plant that size is in the

range of $250 million, and revenue for such a plant is in the range of $2 million per day! (Raw material cost is very small.) The industry standard to build a 6-inch plant was 30 months up to the time the equipment was qualified, that is, no production quantities. The industry standard to get the plant up and running to 90% of capacity is about 46 months. The plant was recently completed and up to 90% production in 13 months. Harris presented their results at a recent conference hosted by the Avraham Y. Goldratt Institute. See their Internet page [15].

▶ *Israeli aircraft industry.* The Israeli aircraft industry employs about 15,000 people. A major function is to maintain jumbo jets used in passenger service. A particular type of maintenance, type D maintenance, normally takes 46 days in the industry. The penalty for nonperformance to schedule is steep—$60,000 per day—because the airlines need the planes back into scheduled service. The company had been paying up to $25 million per year in penalties. A letter from the manager to Dr. Goldratt noted that "we succeeded to drop our average turn around time per aircraft visit from three months to two weeks and to increase our backlog from two months to one year" [16].

▶ *BOS.* According to Izzy Gal, president of Better Online Solutions (BOS), "A project was originally planned to be released to the market in August 1997 (there is no reason to believe that it would have been on time—but who knows?). The TOC scheduling cut four months from this timetable—so it was planned to be ready on May 1, 1997. It was finished in [the] beginning of April, 1997, almost a month before the corrected time. Almost five months before the original time." [17]

1.4   Summary

This chapter defined the problem that this book aims to resolve and identified critical chain project management (CCPM) as a new theory (hypothesis) to resolve the problem. Key points are:

▶ Projects success rate using the existing critical path paradigm is poor for all types of projects in all types of cultures.

▶ Hypothesized causes of project failure do not address the project management system as the potential cause, most often leading to remedies of working harder with the old system. That does not seem to be the right problem.

▶ Evidence suggests that the right problem is in the design of the project system itself; specifically, the system fails to properly manage the reality of uncertainty.

▶ The right solution requires a project system that has a much higher success rate and that is simple to use.

▶ A growing body of evidence does not contradict the hypothesis that Goldratt's critical chain method satisfies the necessary conditions for project success.

Comparing the results of applying the critical chain theory to the existing theory (i.e., the critical path theory as described in the PMBOK™) provides support for using the critical chain theory while we continue to critically review and improve it.

References

[1]   Duncan, W. R., et. al., *A Guide to the Project Management Body of Knowledge*, Upper Darby, PA: Project Management Institute, 1996.

[2]   Goldratt, E. M., *The Goal*, Great Barrington, MA: North River Press, 1984.

[3]   Popper, K. R., *Objective Knowledge: An Evolutionary Approach*, Oxford: Clarendon Press, 1997, p. 144.

[4]   GAO/T-RCED-97-92, "Department of Energy: Improving Management of Major System Acquisitions," *Testimony*, March 6, 1997 .

[5]   GAO/T-NSIAD-97-262, "Space Station: Deteriorating Cost and Schedule Performance Under the Prime Contract," *Testimony*, Sept. 18, 1997.

[6]   Lewis, J. P., *The Project Manager's Desk Reference*, Chicago: Irwin, 1995, p. 245.

[7]   Bromilow, F. J., "Measurement of Scheduling of Construction Time and Cost Performance in the Building Industry," *The Chartered Builder*, Vol. 10, 1974.

[8]   Chun, D. W. M., and M. M. Kummaraswamy, "A Comparative Study of Causes of Time Overruns in Hong Kong Construction Projects, S)263-7863(96)0039-7, *Inter. J. Project Management*, Vol. 15, No. 1, Feb., 1997.

[9]   Leopold, A., *Game Management*, University of Wisconsin Press, 1933.

[10] Goldratt, E. M., *Critical Chain*, Great Barrington, MA: North River Press, 1997.

[11] Lambert, L. R., "Cost/Schedule Control Criteria (C/SCSC): An Integrated Project Management Approach Using Earned Value Techniques," *The AMA Handbook of Project Management*, New York: AMACOM, 1993.

[12] Kahneman, D., P. Dlovic, and A. Tvershky, *Judgment Under Uncertainty: Heuristics and Biases*, Cambridge: Cambridge University Press, 1982.

[13] Goldratt, E., "My Saga to Improve Production," New Haven, CT: Avraham Y. Goldratt Institute, 1994.

[14] "RNLAF Team Seeks Improvement," *Horizons*, Albuquerque, New Mexico: Honeywell Defense Avonics Systems, Vol. 5, No. 2, Feb. 20, 1998.

[15] Rizzo, A., "The TOC Solution of R&D and Multi-Projects Organizations," Whippany, NJ: Lucent Technologies, January 5, 1998.

[16] http//www.tp.semi.harris.com/raptor.html.

[17] http//www.Goldratt.com (Internet site for Avraham Y. Goldratt Institute).

Contents

The synthesis of TQM, TOC, and PMBOK™

This book approaches the problem of improving project management from the perspective of synthesizing three areas of knowledge: PMBOK™, total quality management (TQM), and the theory of constraints (TOC). These three knowledge areas provide different reality filters, or paradigms, for understanding the project system. Three perspectives enable deeper understanding of the theory underlying CCPM. The underlying theory enables you to deal with issues unique to your environment or project.

Figure 2.1 illustrates how the three perspectives on the project system might look at problems in project performance. The PMBOK™ perspective compares actual project system performance to the PMBOK™ model, which it assumes is correct. Therefore, the PMBOK™ perspective is unlikely to blame the PMBOK™ project system as the cause of the problems. It is much more likely

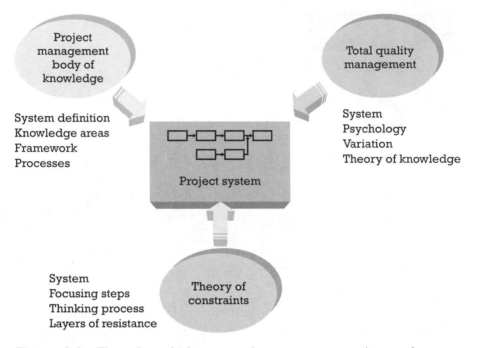

Figure 2.1 Three knowledge areas increase perspective on the project system.

to blame performance problems on failure to properly execute the (assumed) effective system. That is indeed the nature of much of the project management literature. Dr. W. Edwards Deming noted that we should not expect significant system changes to come from within the system. A natural consequence of solutions based on the PMBOK™ perspective is to "do more better."

The TQM perspective continually improves every process. It therefore tacitly assumes that the best way to improve a system is to improve every process. A leading consideration in TQM (profound knowledge) provides four subperspectives that lead to deeper understanding of the potential causes of project problems. TQM provides specific tools to perform root cause analysis to identify the causes of problems and develops strategies to remove those causes.

The TOC perspective identifies the system constraint and works to improve its throughput. It provides a system view of projects and a specific theory to predict project performance and the impact of changes to the system. The TOC perspective differs from the PMBOK™ view by considering the project system as a dynamic process to create completed projects. TOC looks at individual project tasks as the operation of a system for producing the result or output of the tasks. It focuses on the fact that the task performance process includes natural variation and that the individual project tasks are interrelated.

2.1 PMBOK™

Project management made a great leap forward in the 1950s and 1960s with the advent of the CPM and the Program Evaluation and Review Technique (PERT). PERT was developed in 1958 as a joint effort between the United States Navy and the Booz, Allen, Hamilton consulting firm for the Polaris submarine project. The method was enabled by the innovation of computers and was successful in managing the Apollo project to put people on the moon and many large defense projects.

Personal computers have brought sophisticated computer scheduling techniques to everyone's desk. CSCSs have increased the complexity of these systems. However, there has been little progress in improving the success rate of projects and even less innovation in the underlying basis and system. People continue to work with project management assumptions conceived 40 years ago.

Figure 2.2 illustrates the related knowledge areas identified in the PMBOK™ Guide. This text focuses on and proposes changes to the project management knowledge elements to affect the necessary conditions for project success: project integration management, project scope management, project time management, and project risk management. You must address the other knowledge areas to varying degrees, depending on your projects and the environment in which you work.

The PMBOK™ Guide describes general processes for each knowledge area, collected into five types of processes:

- Initiating;

- Planning;

- Controlling;

- Executing;

- Closing.

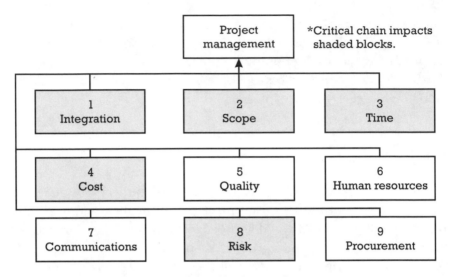

Figure 2.2 PMBOK™ areas identify the project system.

These process phases roughly correspond to the phases of a project, but there is considerable overlap. The PMBOK™ Guide emphasizes that there are relationships and interactions between most of the project system processes.

Perhaps the most important element of successful projects is the project team. An able leader and an effective team can achieve project success despite a flawed system. A weak team or leader will struggle with even an excellent system. While the human side of project management is extremely important to project success, it is not a specific topic of this book.

2.1.1 Project integration management

Project integration management includes project plan development and execution and overall change control through the life of the project.

2.1.2 Project scope management

Project scope management includes the process leading to initiation of the project and scope planning, definition, verification, and change control. Primary outputs of the scope management processes include a project charter, the project work breakdown structure (WBS), detailed statements of work (SOWs), functional and operational requirements (F&OR) or other definitions of the deliverable scope, the project assumptions, and a process for scope change control.

Project assumptions help planners develop a deterministic project plan. The plan and control processes defined by the PMBOK™ do not include a way to handle decision branches in a project plan. Assumptions define uncertainty sufficiently to permit defining a deterministic scope, cost, and schedule. For example, some of the research projects I have worked on required unprecedented performance of technical parts. If the parts succeeded in delivering the specified performance, we would follow the path of building the units for installation in a larger system (in this case, a nuclear reactor). If the part did not perform as specified, we had to modify the design and test again. Yet I could have only one project plan. I usually assumed we got it right the first time, even though I knew we would not always succeed. That permitted building a plan and a cost estimate. I then addressed the potential of failure to succeed as part of the project risk management.

2.1.3 Project time management

Project time management includes defining the activities necessary to produce the project scope, sequencing the activities, estimating activity duration, developing the project schedule, and controlling the project to the schedule. Schedule preparation requires the WBS and scope statements as inputs. The schedule development process identifies the activity resource requirements and other potential project constraints. The guide also discusses the need to level resources in the plan, that is, to adjust the planned resource demand of the plan to match the expected resource supply.

The PMBOK™ Guide notes that activity duration estimates should specify uncertainty and refers the reader to discussions on project risk management to handle that uncertainty. It does not differentiate between common cause variation and special cause variation. Thus, it includes all potential variation into the single category of project schedule risk management.

The PMBOK™ Guide addresses cost management as a separate topic from time management, but the processes are identical. The schedule and cost control process includes updating the schedule and completng the project estimate, planning and executing corrective action, and assessing the lessons learned at the close of the project.

2.1.4 Project risk management

Project risk management includes identifying and quantifying risks and planning and controlling response to risk. Risk includes both the likelihood and the consequences of adverse impacts to the project. The PMBOK™ Guide does not distinguish between common cause and special cause variation but appears to lump them together in the performance of risk management.

2.1.5 Other PMBOK™ areas

The other PMBOK™ knowledge areas, including quality, human resources, communications, and procurement management, are all important to projects. They are important to any type of business. The scope of this text does not explore those areas.

2.2 TQM

The popular literature may lead you to believe that TQM was a management fad that failed to deliver on its promise and had outrun its applicability by the end of the century. Nothing could be further from the truth. At the February 1999 award ceremony in Washington, D.C., President Clinton noted that previous winners of the national Malcolm Baldrige quality award from 1988–1997 posted an impressive 460% return on investment, compared to a 175% increase for the S&P 500 over the same period. Hendricks and Singhal published results in April 1999 demonstrating performance measures for TQM award-winning firms outstripping comparison control firms by two to one [1]. For example, the TQM firms posted a 91% (vs. 43% for non-TQM firms) increase in operating income, a 69% (vs. 32%) increase in sales, and a 79% (vs. 37%) increase in total assets.

Dr. W. Edwards Deming, the man most people consider the father of TQM, never defined TQM. Deming described his approach in seminars and books [2,3], and though a great advocate of operational definitions, he chose to never offer one for TQM. Instead, he preferred to discuss the matter in terms of his 14 points, or "Principles for the Transformation of Western Management." He supplemented those points with identified diseases and obstacles to achieving the transformation he preached.

Dr. Deming's 14 points for management [2] are:

1. Create constancy of purpose toward improvement of product and service, with the aim to become competitive and to stay in business, and to provide jobs.

2. Adopt the new philosophy. We are in a new economic age. Western management must awaken to the challenge, must learn their responsibilities, and take on leadership for change.

3. Cease dependence on inspection to achieve quality. Eliminate the need for inspection on a mass basis by building quality into the product in the first place.

4. End the practice of awarding business on the basis of price tag. Instead, minimize total cost. Move toward a single supplier for any one item, on a long-term relationship of loyalty and trust.

5. Improve constantly and forever the system of production and service, to improve quality and productivity, and thus constantly decrease cost.

6. Institute training on the job.

7. Institute leadership (see point 12). The aim of supervision should be to help people and machines and gadgets to do a better job. Supervision of management is in need of overhaul, as well as supervision of production workers.

8. Drive out fear, so that everyone may work effectively for the company.

9. Break down barriers between departments. People in research, design, sales, and production must work as a team, to foresee problems of production and in use that may be encountered with the product or service.

10. Eliminate slogans, exhortations, and targets for the work force asking for zero defects and new levels of productivity. Such exhortations only create adversarial relationships, as the bulk of

the causes of low quality and of low productivity belong to the system and thus lie beyond the power of the work force.

11. (a) Eliminate work standards (quotas) on the factory floor. Substitute leadership. (b) Eliminate management by objective. Eliminate management by numbers, numerical goals. Substitute leadership.

12. (a) Remove barriers that rob hourly workers of their right to pride in workmanship. The responsibility of supervisors must be changed from sheer numbers to quality. (b) Remove barriers that rob people in management and in engineering of their right to pride of workmanship. This means, inter alia, abolishment of the annual or merit rating and of management by objectives.

13. Institute a vigorous program of education and self-improvement.

14. Put everybody in the company to work to accomplish the transformation. The transformation is everybody's job.

In later life, Deming brought together the overall methods he believed in under the title of "profound knowledge" [3]. Deming defined a system of profound knowledge as a lens and a map of the theory to understand and optimize organizations. He emphasized that profound knowledge is itself a system, having an aim and with all the parts interconnected. He identified four segments for discussion but emphasized that they cannot be separated. The four elements are:

▶ Appreciation for a system;

▶ Knowledge about variation;

▶ Theory of knowledge;

▶ Psychology.

Figure 2.3 illustrates that those four ideas are interrelated. The following section discusses each element as it relates to the project management system. Subsequent sections describe these areas, emphasizing the relationship to the project system.

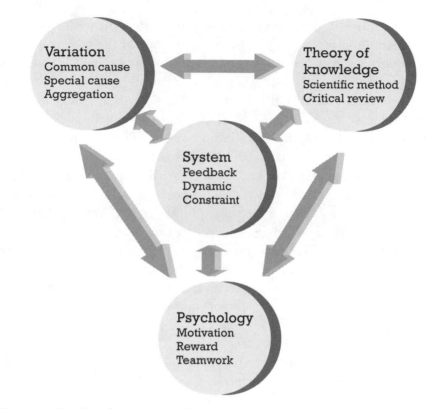

Figure 2.3 The four areas of profound knowledge are interrelated.

2.2.1 Appreciation for a system

Every system must have a defined aim or goal. That is the purpose of the system and defines the boundary of the system. The system itself is a network of interdependent components that work together to try to accomplish the aim of the system. Profit-making business systems have the goal of making money, now and in the future. That is why people invest in profit-making businesses. Nonprofit businesses (those intended to be that way, anyhow) have different goals, for example, creating health for a health care institution or creating family well-being for some social institutions. Projects have the goal of delivering to the customer a specified unique product or service on time and within cost. The client for that goal can relate the project result to the broader goal of the institution.

The project system consists of physical things, people, and nonphysical things, such as policies, knowledge, and relationships. All those things are interconnected to varying degrees and may affect the performance of the system. Project planning and control are part of the project system. Task performance by the project team is part of the project system.

Things outside the system may affect it. Business systems are open systems, which means that energy and physical things flow through them. Project systems are the same. These things flowing through, such as people, policies, and capital, can affect the system. For example, laws and regulations, which are outside both the business system and the project system, can have an immense impact on the performance of the system.

In 1950, Dr. Deming drew a sketch similar to Figure 2.4 on a blackboard in Japan. He attributes the subsequent success of postwar Japan in large part to the understanding conveyed by the figure. His description of the system starts with ideas about possible products or services. He considers these ideas predictions of what the customer might want or need. The prediction leads to the decision to design the product or service and to test it in preliminary trials before committing to full-scale production. Feedback from the customers is a key part of driving the system toward the future.

The project management system operates in precisely the same way when you consider it delivering project after project. Customers specify what they want from the project. The project team prepares a project plan to create the specified result. The plan brings together various functions within the company and purchased services and parts to produce the desired result. Just as a company may produce many products or deliver many services, the project management system is capable of delivering many completed projects. Although the deliverables from specific

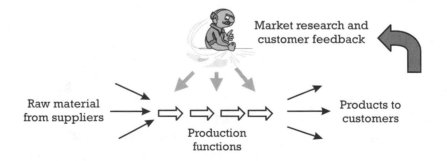

Figure 2.4 Deming's sketch of a business system emphasized interrelationships and feedback.

projects are unique, the same project management system serves to produce the results.

Deming understood system dynamics. He observed that operation of his flow diagram required the flow of material and information from any part of the system to match the input required by the next element in the system. He emphasized that the definition of the system must consider the impact on the future of the system. He makes reference to the following material.

2.2.1.1 System dynamics

Senge [4] described the essence of the discipline of systems thinking as a shift of the mind to:

▶ Seeing interrelationships rather than linear cause-effect chains;

▶ Seeing processes of change rather than snapshots.

He presented the "laws of the fifth discipline" to summarize and understand how dynamic systems (including a project system) work. The following list gives the laws and an instance of how each applies to the project management system.

1. Today's problems come from yesterday's "solutions." (Management was unsatisfied with too long a schedule on the last project, so they cut the individual task estimates. This time, people added in a margin for management to cut out.)

2. The harder you push, the harder the system pushes back. (Management works to increase efficiency by assuring that all resources have multiple project tasks on which they can work. Working on multiple tasks means all tasks take longer, since people can really only work on one task at a time. The others are doing nothing while one is worked on. Projects get longer and longer and efficiency goes down.)

3. Behavior grows better before it grows worse. (Management puts selected resources on overtime to accelerate the project. Results improve. The resources then get used to the extra income and slow down to not work themselves out of a job.)

4. The easy way out usually leads back in. (The *Mythical Man Month* [5] explains this in detail. Management adds resources to recover the schedule on a project that is slipping. Management must search for the people, hire them, create places for them to work, purchase them tools, and integrate them into the project team. The last step, in particular, requires the time of the most productive project resources. The project falls further behind.)

5. The cure can be worse than the disease. (The most common solution to improve project performance is to use more rigor and make more detailed project plans. This often helps on a project just after a major project disaster, due to regression to the mean. That is, it is unlikely that two projects in a row will have a bunch of bad breaks. So from then on, project plans are more complex and require more paperwork. Attention moves from completing the project tasks to completing the paperwork. Project durations and costs increase. Project changes increase, further increasing cost and time.)

6. Faster is slower. (The team passes on a piece of software that really needed two more days of testing to meet their milestone date. The software causes problems in the integrated system test, which takes weeks to diagnose.)

7. Cause and effect are not closely related in time and space. (The space shuttle blows up on launch from Cape Kennedy in Florida. The cause is a seal design made and tested in Utah years before but not previously subjected to specific environmental conditions. The Hubble space telescope is near-sighted (a billion-dollar mistake) because crucial testing was skipped years before, on earth, to keep the schedule.

8. Small changes can produce big results, but the areas of highest leverage are often the least obvious. (A major lever for systems containing people is the measurement and reward system. The impacts of measures and rewards are not always well thought out. For example, as Deming notes, monthly quotas lead to the end-of-the-month syndrome, when a lot of bad product is shipped. In projects, management emphasis that people keep to

their commitments causes them to add time to their delivery
estimates and withhold work that is completed early.)

9. You can have your cake and eat it too—but not at once. (The critical chain multiproject process completes more projects much
faster; individual project duration decreases, and the number of
projects complete at any time increases. But you must delay the
start of projects to get the benefit.)

10. Dividing an elephant in half does not produce two small elephants. (Senge relates the tale of the blind men describing an
elephant based on feeling its different parts: the trunk, the
massive body, a leg, and the tail. Of course, their descriptions
vary. Project failure analysis often examines subprocesses, such
as the workplan process or the change control process, to see
what part of the system needs to be repaired. That approach
fails to examine the underlying assumptions, for example, the
assumption of deterministic task schedules implicit to printing
out start and finish dates for thousands of tasks.)

2.2.1.2 Leverage

Dynamic systems lead logically to consideration of the possibility of using
the system itself, as in jujitsu, to move the system in the direction you
want it to go. Leverage defines small changes (inputs to the system) that
cause large results (outputs from the system). The idea is like compound
interest: if given enough time to work, a small interest rate can lead to
very large accumulation of capital. People who knowingly work with
complex systems focus on trying to find high-leverage interventions to
cause desired outcomes.

Senge noted that there are no simple rules to find high-leverage
changes to improve systems, but that thinking about the underlying system structure, rather than focusing on events, makes finding those
changes more likely. I contend that the major reason that there has not
been a significant improvement in project management prior to CCPM is
that all the observers were looking at the problem from the same flawed
perspective; that is, not looking at it as a system comprising people, things,
and information. They appear to have asked, "How do I operate this system better?" They should have asked, "How do I improve this system?"

Due to the effect of compounding, it is likely that any high-leverage interventions in systems will be in the feedback loops. Feedback loops affect the system based on the results that obtain. More results cause more feedback, so such loops are similar in impact to compounding interest. Powerful feedback loops for systems involving people always include the measurement and reward systems. Thus, the project performance measurement system is one area we can focus on to leverage improvements in project success.

2.2.1.3   Unintended consequences

The linkage and correlation between the parts of a system mean that changing any part of the system may influence other parts. As noted in the laws of the fifth discipline, the change may be in a desired direction or not, it may be large or small, and it most likely will not be at the same time and in the same place that caused the change. Many people, especially those prone to fiddling with social systems, talk in terms of "unintended consequences." Hardin [6] made the point that, from the ecological view of systems, there is no such thing as unintended consequences. When you change part of a system, other parts change. That is it! You can count on some of those changes as being undesirable to one or more perspectives of the system. Therefore, you must use caution when posing changes to a system such as the project management system. Some of the changes posed to eliminate certain undesired results or root causes may have worse consequences elsewhere.

Several aspects of the project system illustrate that effect. For example, if we make the consequences for delivering a task result on a project severe, it is likely to cause all subsequent estimates to include additional contingency. It may cause quality of output to go down, influencing other tasks later in the project.

2.2.1.4   Destruction of a system

Destruction of a system by forces within the system was one of the key issues that Deming tried to bring home to management. He discussed how selfish competition versus cooperation between departments often causes such destruction. Senge [4] and Deming [2] illustrated numerous examples of how government attempts to improve things often lead to destruction of the very system they were hoping to improve.

In project systems, conflicts may arise between the client and the project team, between senior management and the project team, between different parts of the project team, between the project team and supporting organizations within the company. A frequent example of conflict between the project team and supporting organizations is the nearly continual battle between procurement organizations and project organizations in large companies; especially those doing work for the federal government. Often the procurement organization's primary measures relate to compliance with a complex system of procurement regulations and policies, while the project team is interested only in having it fast and good. Sometimes the procurement organization's goal is to get it cheap, while the project organization wants it good. The project system design must ensure that the measures and rewards of individual parts of the organization cause the parts to work together to support the whole. Deming notes: "The obligation of any component [of a system] is to contribute its best to the system, not to maximize its own production, profit, sales, nor any other competitive measure." [2]

2.2.2 Understanding variation and uncertainty

> I returned, and saw that under the sun, that the race is not to the swift,
> nor the battle to the strong, neither yet riches to men of understanding,
> nor yet favor to men of skill; but time and chance happeneth to them all.
>
> *(Ecclesiastes 9:11)*

A project system attempts to predict and produce a certain result for a certain cost by a certain time. As the quote from Ecclesiastes illustrates, people know full well that the world is an uncertain place. Variation exists everywhere. Predictions are never completely accurate.

Understanding variation is essential to making any real system operate. Popper, in an essay titled "Of Clouds and Clocks," described a range of reality fundamental to understanding variation [7]. He bids us to consider a horizontal line, with a clock on the right representing the ultimate of a clockwork-like deterministic world. In that world, everything would eventually be completely predictable; it is only a matter of understanding completely the cause-effect relationships that determine the workings of this mechanical model. The ultimate manifestation of this model is the working of the planets of the solar system, whose motions are predictable with uncanny accuracy using the equations defined by Isaac Newton.

The cloud, at the other extreme of Popper's continuum, represents complete chaos—not the deterministic chaos of current mathematics, but the random chaos associated with the world of complete uncertainty. It represents the unpredictability of science at the quantum level and the unpredictability of nature at the human scale. Popper wrote, "My clouds are intended to represent physical systems which, like gases, are highly irregular, disorderly, and more or less unpredictable." Everything falls between those two extremes.

Uncertainty means indefinite, indeterminate, and not certain to occur, problematical, not known beyond doubt, or not constant. All predictions are uncertain. Fundamental physics tells us that all knowledge of reality is uncertain; the better we know the position of something, the less we know about how fast it is moving. Uncertainty is the true state of the world.

Project managers can predict many things well enough to achieve the things they plan, such as building a house. Scientists also know that we can never accurately predict certain other things. For example, no matter how well we learn to model the weather, and how well we measure conditions at one point in time to run the model, our ability to predict specific phenomena will always be limited by the nature of the physical laws that determine local weather behavior. Scientists now know (from the chaos theory) that they will never be able to predict when and where the next tornado will touch down. On the other hand, they can predict seasonal trends reasonably well.

Starting in the seventeenth century, mathematicians and scientists have sought to improve the ability to predict the world further and further over into the cloudy region. At the same time, science kept moving the cloudy region to include more and more of nature. It extended the smallest scale with quantum mechanics, and showed cloudiness at the largest scale with increasing understanding of the universe. Cloudiness encompassed all intermediate scales with the discovery of chaos and study of complex adaptive systems.

2.2.2.1 Common and special cause variation

Probability and statistics are science's weapons of choice to deal with cloudy systems. Shewhart [8], a mentor to Dr. Deming, identified the need to operate systems in a state of statistical control to have a degree of predictability. He observed, "Every mathematical theorem involving this mathematically undefined concept [statistical control] can then be given the following predictive form: If you do so and so, then such and such will happen."

Following Shewhart, Deming emphasized the importance of distinguishing between common cause variation and special cause variation. It is necessary to distinguish between them to get a system under statistical control. It is necessary to have a system under statistical control to predict its future performance. Common cause variation is variation within the capability of a system to repeatedly produce results. Special cause variation is variation beyond that range; usually variation with causes outside the system. Management's function is to improve the system while avoiding two mistakes:

> Mistake 1: Treating common cause variation as if it were special cause variation;

> Mistake 2: Treating special cause variation as if it were common cause variation.

Dr. Deming called mistake 1 "tampering." Tampering is making changes within a system that is operating in statistical control. Tampering always degrades the performance of a system. He described the case of a machine that had a feedback device attached to measure each part and to adjust the tool location based on that measure to try to improve the repeatability of each part. It made the variation in parts much larger, because the measurements included the natural variation (capability) of the system to produce parts. The tool simply amplified that natural variation.

Tampering relates to the measurement and control of project performance, and the decisions to take management actions based on those measurements. This phenomenon means that responding to common cause variation as if it were special cause variation will make the system performance worse. In other words, responding to small variances by making project changes degrades project performance.

The government provides an ongoing set of examples for mistake number 2. Something undesirable happens, and they put in place a law to ensure it never happens again. We end up with thousands of pages of regulations and laws, each applicable to some rare event or events not even applicable to the subject of the action. Mistake 2 is the essence of the growth of bureaucracy. It happens in business every bit as much as in government.

All the estimates in a project plan are uncertain. Performing each of the tasks within a project plan is a single trial of a system (the project task performance system) and is, therefore, unpredictable. However, statistical techniques enable us to predict with known precision the likely results of numerous trials from a production system and to separate out the special causes of variation requiring corrective action. While knowledge of variation has been used to great profit in production operations, it has not (until now) been used to improve project performance. The PMBOK™ Guide and the supporting literature we have examined fail to differentiate between common cause variation and special cause variation, a major oversight in the current theory.

2.2.3 Psychology

Several properties of the human mind lead to individual behavior that seems to resist change. B. F. Skinner described one of the more powerful mechanisms [9]. Skinner asserted (with extensive scientific data) that much human behavior comes from he called operant conditioning. Put simply, that means people continue to do what gives positive reinforcement and learn to avoid doing things that do not lead to positive reinforcement or that help them avoid negative reinforcement. Positive reinforcement is something a person likes. Negative reinforcement is something a person does not like. Positive and negative reinforcers vary from individual to individual. Skinner noted, "A reinforcing connection need not be obvious to the individual reinforced."

Figure 2.5 illustrates my rendition of a control system view of Skinner's model. It starts with a need, which is influenced by the person's current state, including deprivation or satiation relative to the goal. Comparing that need to the person's understanding of his or her current situation (perceived reality) yields a gap that, if large enough, motivates the person to action. Action seeks to change reality to close the gap. The sensor, which may be the five senses or more removed methods of gaining data, feeds back information about the effect that the action has on reality. If the change is positive (reducing the gap or otherwise

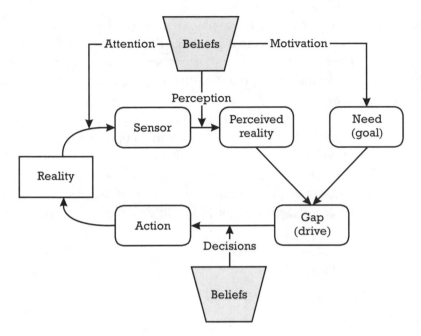

Figure 2.5   Control system view of human actions (behavior).

supplying a reward), it strengthens the chances that the person will repeat the behavior. That is what Skinner called operant conditioning.

Operant conditioning must be somehow stored in a person's brain. Because it defines a (perhaps rudimentary) model of the world (if I do this, then I get that), it can be considered a belief about how the world works. Such beliefs may be conscious or unconscious. Research demonstrates that these beliefs have other impacts on the model. Figure 2.5 illustrates that beliefs affect what we pay attention to, how we interpret what we sense (perception), what our motivations (needs) are, and the decisions we make on how to act in the world so as to increase rewards and decrease negative reinforcers. The influence is mostly unconscious. In other words, we see it because we believe it.

2.2.3.1 Rewards

While operant conditioning works well for rats and pigeons, extreme care must be used in applying the model to human beings. Much of the damage done in organizations follows directly from applying oversimplified models of operant conditioning to human beings. The field of performance measurement and concepts such as pay for performance are just some of the worst examples of ineffective practices derived from oversimplified application of the reward-punishment concepts, even though Skinner identified and described in depth and proved by experiment that punishment does not work.

Worse yet, research with humans conclusively demonstrates, time and again, that rewards work only to motivate people to get the reward. Usually there are more unintended negative consequences from reward systems than positive benefits. Kohn describes the reasons for this, noting that reward and punishment are really two aspects of the same thing: attempts at external control [10]. He explains five reasons why rewards fail:

1. Rewards punish;

2. Rewards rupture relationships;

3. Rewards ignore reasons (for the problem that elicited the need for a reward);

4. Rewards discourage risk taking;

5. Rewards cause people to lose interest in the task itself and therefore to lose intrinsic motivation.

That is not news, but much of modern management does not get it. Frederick Herzberg noted:

> Managers do not motivate employees by giving them higher wages, more benefits, or new status symbols. Rather, employees are motivated by their own inherent need to succeed at a challenging task. The manager's job, then, is not to motivate people to get them to achieve; instead, the manager should provide opportunities for people to achieve so they will become motivated. [11]

The requirements for CCPM must include designing the system to provide such opportunities. A significant barrier in the deterministic critical path approach is that workers win or lose depending on whether they complete their tasks on time. Yet all involved know full well that the task duration estimates in the schedule have significant uncertainty. As

Dr. Deming demonstrates with his bead experiment [2], random fluctuations determine employee success or failure. That system clearly does not meet the design requirement.

2.2.3.2 Additional psychological considerations

The modern view is that our minds operate as pattern recognition devices. We have a wonderful ability to infer the automobile in the picture by looking at only a small fragment of the picture. We often can name that tune in three notes. It is remarkable, when you think about it.

Beliefs act to focus our attention, and they adjust our perception of reality by acting as a kind of information filter. Two people witnessing the same events may have dramatically different views of what happened. I was fascinated while listening to congressmembers from both parties arguing the impeachment of President Clinton. Participants from both sides made logical and emotional arguments for their positions. No one argued that they held their position because of the political party with which they were aligned. Yet, when the vote came in, only five representatives of 417 crossed the party line in their vote. While I am certain that a small minority literally chose to vote with the party, the speakers convinced me that they really believed the logical arguments that they made for their side. Because the argument was framed as an either-or choice, one would expect that arguments based on factual analysis should have aligned people regardless of political party. My perceptive filter saw that as an outstanding example of how people interpret the facts (i.e., perceive) in ways that align reality with their beliefs. The impassioned logical arguments of both sides had no impact whatsoever on the other side, because they did not change the basic underlying beliefs. The participants in the debate were locked into their own paradigms.

At any point in time, people operating in an environment tune their behavior to the environment. Put another way, feedback through operant conditioning causes them to behave in ways that maximize positive reinforcement and minimize negative reinforcement in the current environment. Changes in the system threaten that position. Furthermore, Skinner demonstrated that extinguishing behavior established by operant conditioning can take a long time. The organism will continue to emit the old behavior, which is no longer reinforced, sometimes for thousands of tries.

Other aspects of psychology, or how our minds work, are also important to understand the system you are attempting to change. One of those aspects is the availability bias. Psychological experiments repeatedly demonstrate that we are relatively poor judges of probability. Instead, we focus on the information we heard or saw last or that impressed us the most, when offering judgments about probability. For example, you will often hear statements such as, "All scientists (programmers, engineers, etc.) tend to underestimate how long it will take to do a task." When pressed for data, people admit to having little. Data analyses often prove otherwise. Most project tasks are reported as complete on the due date. (A miraculous occurrence, by the way, proving the existence of date-driven behavior.) People also tend to be overconfident in their ability to estimate probabilities.

The PMBOK™ Guide does not deal with psychology as a knowledge area. Despite that, many project management texts deal with the human side of project management. The project system must integrate with the human subsystem. The integration happens through the psychology of individuals and groups. Because the current system was not designed with that connection in the forefront, you may expect to find some problems in this area. Chapter 3 demonstrates that the core conflict leading to most of the observed undesired effects with the current project system stem from a mismatch between individual psychology and the project system goal.

2.2.4 Theory of knowledge

Popper, in an essay titled *Conjectural Knowledge*, stated, "From a rational point of view, we should not rely on any theory, for no theory has been shown true, nor can be shown to be true" [7]. That point, agreed on by most philosophers and scientists, is far from the understanding of the common person, who is prone to accepting a single instance that conforms to a theory as evidence that the theory is right. Popper went on to state:

> In other words, there is no "absolute reliance"; but since we *have* to choose, it will be "rational" to choose the best tested theory. This will be "rational" in the most obvious sense of the word known to me: the best tested theory is the one which, in the light of our *critical discussion*, appears to be the best so far, and I do not know of anything more "rational" than a well-conducted critical discussion. [7]

Popper also suggested an objective criterion to prefer a new theory, "is that the new theory, although it has to explain what the old theory explained, *corrects* the old theory, so that it actually *contradicts* the old theory: it contains the old theory, *but only as an approximation.*"

Figure 2.6 illustrates the scientific method. The method operates based on effect→cause→effect. Scientists start by defining a problem: hypothesizing the cause for an observed effect. All new theories have some confirming evidence; that is why a scientist proposed the new theory. The prediction of a previously unseen effect that differentiates the new theory from the old tests the theory. Existence of the predicted effect provides evidence to prefer the new theory to the old. Lack of the effect fails to provide evidence to prefer the new theory. A theory is usable until disproved. A successful experiment does not mean that it is correct (true), and it does not mean that it will work in the future. A successful experiment just means that it worked over the domain so far experienced.

A commonly used example of the scientific method is Newton's laws of motion and gravitation. Before Newton, many data were gathered on the positions of the sun and the planets. Correlations were used to develop quite accurate predictions of the motion. There was a fundamental flaw, of course, in that they had the earth at the center of the solar system. Nevertheless, the correlations worked.

Newton's laws worked better than those of his predecessors because they extended beyond what had been observed. Newton's laws allowed

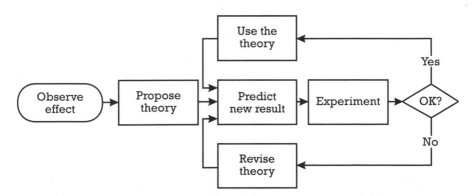

Figure 2.6   The scientific method checks the validity of a theory by experiment. No theory is ever proven. It is accepted as "good enough" to use until it is rejected by a single experiment or replaced by a theory that better predicts reality.

prediction beyond the realm of the observed and allowed us to put humans on the moon and to send spacecraft to Jupiter. That is impossible using correlation of planetary movement.

Then along came Albert Einstein. His equations proved that Newton's equations are wrong. (Newton knew that; he had proposed them as "good enough.") Einstein's equations reduced Newton's equations where speeds are modest compared to the speed of light and where gravity is not too large. That fits Popper's model of a better theory. Einstein spent his later life trying to prove his own theory wrong by developing a unified theory. So far, no theory better than Einstein's theory has been found. Therefore, scientists continue to use Einstein's theory. This is a theory of knowledge at work.

Understanding the theory of knowledge enables you to better test the CCPM theory compared to the critical path theory or other theory of project management you are currently using. You now know you can never prove a theory true, but you have working tools (test and critical discussion) to choose between competing theories. The theory of knowledge will also help you make decisions necessary to plan a specific project and to operate the project system you choose.

2.3 TOC

Basically, TOC is a commonsense way to understand a system. TOC says, "Any system must have a constraint that limits its output." We can prove that with critical discussion. If there were no constraint, system output would either rise indefinitely or go to zero. Therefore, a constraint limits any system with a nonzero output. Figure 2.7 shows that limiting the flow through any of the arrows can limit the total output of the system. That

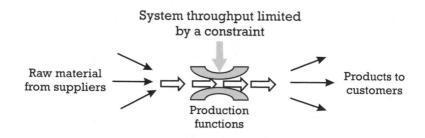

Figure 2.7   TOC limits the output of a system by a constraint.

arrow would be the system constraint. People identify the constraint in physical systems as a bottleneck, a constriction that limits flow through the system.

The purpose of using TOC is to improve a business system. In *What Is This Thing Called Theory of Constraints, and How Should It Be Implemented?*, Goldratt stated:

> … before we can deal with the improvement of any section of a system, we must first define the system's global goal; and the measurements that will enable us to judge the impact of any subsystem and any local decision, on this global goal. [12]

Dr. W. Edwards Deming noted in *The New Economics* that "We learned that optimization is a process of orchestrating the efforts of all components toward achievement of the stated aim" [2].

A physical chain provides the most commonly used prop to describe TOC. The goal of a chain is to provide strength in tension. Everyone accepts that the weakest link determines the strength of a chain. Anyone can see that improving the strength of links other than the weakest link has no impact on the strength of the chain (see Figure 2.8).

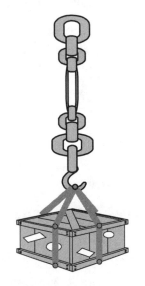

Figure 2.8   A physical chain illustrates TOC in action: The weakest link constrains the strength of the chain.

The next step in understanding TOC is not so evident. TOC makes a leap to throughput chains and poses the theory that for any chain, throughput (at any time) is limited by, at most, one constraint. Perhaps that is easier to see in the project world, where a project plan can have only one longest path. The only case that would have that as not true is if two or more paths are exactly the same length. As soon as you start to perform the project, it is likely that one path will become the real constraint. The constraint (the longest path) will seem to shift due to fluctuations in project activity performance. But at any time, only one controls the actual time to complete the project.

Applying the scientific method to this basic understanding of TOC leads to many principles. William Dettmer posed the following list [13]:

1. System thinking is preferable to analytical thinking in managing change and solving problems.

2. An optimal system solution deteriorates after time as the system's environment changes. A process of ongoing improvement is required to update and maintain the effectiveness of a solution.

3. If a system is performing as well as it can, not more than one of its component parts will be. If all parts are performing as well as they can, the system as a whole will not be. *The system optimum is not the sum of the local optima.*

4. Systems are analogous to chains. Each system has a weakest link (constraint) that ultimately limits the success of the entire system.

5. Strengthening any link in the chain other than the weakest one does *nothing* to improve the strength of the whole chain.

6. Knowing what to change requires a thorough understanding of the system's current reality, its goal, and the magnitude and direction of the difference between the two.

7. Most of the undesirable effects within a system are caused by a few core problems.

8. Core problems are almost never superficially apparent. They manifest themselves through a number of undesirable effects (UDEs) linked by a network of effect→cause→effect.

9. Elimination of individual undesired effects gives a false sense of security while ignoring the underlying core problem. Solutions that do this are likely to be short-lived. Solution of a core problem simultaneously eliminates all the resulting UDEs.

10. Core problems are usually perpetuated by a hidden or underlying conflict. Solution of core problems requires challenging the assumptions underlying the conflict and invalidating at least one.

11. System constraints can be either physical or policy. Physical constraints are relatively easy to identify and simple to eliminate. Policy constraints are usually more difficult to identify and eliminate, but they normally result in a larger degree of system improvement than the elimination of a physical constraint.

12. Inertia is the worst enemy of a process of ongoing improvement. Solutions tend to assume a mass of their own, which resists further change.

13. Ideas are not solutions.

TOC is a relatively young theory and one that is undergoing continuous improvement. A few years ago, the method to locate what to change in a system relied on discovering the core problem, as illustrated by the above list. The core problem is a problem that, if removed, would begin to cause the system to change undesired effects into desired effects. In other fields, it is called the root cause. There has been a gradual shift to define a core conflict instead of a core problem. This significant step in the theory claims that most of the undesired effects in a system flow from an unresolved, or at least unsatisfactorily resolved, conflict or dilemma. Substituting the term *core conflict* for *core problem* into the preceding list (except for item 10) makes it reflect current understanding. Item 10 in the list was the earlier statement of the current understanding.

The idea of a core conflict underlying system undesired effects must rest on the thought that people would change the system to eliminate undesired effects if they knew how and if they were able to make the changes. If undesired effects persist in a system, something is preventing the system designers or operators from changing the system to eliminate the undesired effect. The core conflict idea helps to identify that something.

2.3.1 The throughput world

Dr. Goldratt found that, most of the time, system constraints trace back to a flawed policy, rather than to a physical constraint. In *The Goal* [14], he demonstrated that policy constraints derived from a flawed system of accounting. Accounting systems in use today were developed around the turn of the century and have changed little since (twice the history of project management systems). When they were developed, they were based on assumptions (no longer listed) about the design of business enterprises.

Dr. Goldratt defined the old accounting system as the "cost world," because it operates on the assumption that product cost is the primary way to understand value and make business decisions. That requires the allocation of many expenses to products, through elaborate product-cost schemes, like activity-based costing. Such schemes are full of assumptions, and often lead to erroneous understanding and decisions.

Dr. Goldratt defined a new way of accounting, which he called the "throughput world." It rests on three definitions:

- *Throughput*: All the money made from selling a product (revenue minus raw material cost);

- *Inventory*: All the money tied up in fixed assets to enable the throughput (the primary difference here is that fixed assets and inventory are treated the same);

- *Operating expense*: All the money spent to produce the throughput.

Major accounting authorities around the world have endorsed this method, but it has been slow to penetrate the mainstream.

The cost world was not bad when it was developed, around the turn of the twentieth century. At that time, big business (which designed it) consisted primarily of plants with very large capital investments, for example, resource industries, steel, railroads, and, a little later, automobile manufacturing, representing fixed cost. At that time, things were tough for labor, which was a variable cost. Labor was mostly applied to unskilled jobs and was plentiful and easy to replace. Therefore, it was easy to vary the workforce with demand.

Today, the skilled workforce is much less variable, and the traditional fixed costs are much less fixed. The concept of allocating costs to labor or products always requires many arbitrary assumptions. Those assumptions, often long forgotten, influence the business decisions made using the cost accounting practices.

The throughput world corrects those errors and focuses all decisions on the goal of the company, that is, to make money now and in the future. All decisions and measures relate to the global goal and often lead to different decisions than those dictated by the cost world.

For example, in the cost world, managers measure operating efficiencies of local workstations. Financial people count inventory as a company asset. If they do not need workers to produce product for customer need, then they produce product for inventory, increasing efficiency to make themselves and their local plant look good. Unfortunately, the plant does not make money on inventory. Inventory costs money to make (raw materials) and to store. So it hurts cash flow and reduces disposable cash at the plant. Our accounting system says it is good, but it is bad for business.

Then, when you get around to selling the inventory (which is good), it reduces your assets (which looks bad). That does not seem to make much sense to me.

On the other hand, what is normally considered the biggest competitive edge in knowledge industries? People. What are people on the accounting system? Expenses. They look bad. They are the first things you want to get rid of if business looks bad; keep the assets, drop the expenses. Dump your ability to make money now and in the future, keep your hardware, which costs you money.

An effective way to evaluate the meaning of the dilemma facing managers is to apply one of the thinking process tools invented by Dr. Goldratt: the "evaporating cloud." Figure 2.9 illustrates the throughput world–cost world evaporating cloud. Block A represents a common objective all managers share. Blocks B and C are requirements to achieve that objective. You read the cloud "To manage properly, managers must control cost." Then you read the lower branch, "To manage properly, we must protect throughput." So far so good.

Focus on throughput requires understanding and controlling the whole system to optimize throughput. The most important effect of throughput world thinking is that it requires focus on throughput as the much preferred path to system improvement. Looking at how throughput, inventory, and operating expense affect net profit and ROI leads to

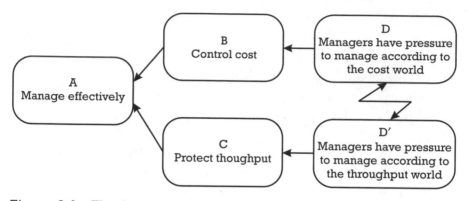

Figure 2.9 The throughput world–cost world evaporating cloud exposes the manager's dilemma.

an immediate conclusion that throughput is the most important variable. Improvements in throughput are unbounded, while improvements in operating expense and inventory (or investment) are limited.

Cost world thinking leads to a piecemeal view of each part of the production system. Costs add algebraically. The cost world leads to focus on operating expense. You can reduce operating expense in any part of the system, and the operating expense reductions add up. This thinking leads to entity D, with the logic, "To control cost, managers have pressure to manage according to the cost world." That is, managers are forced to downsize and cut costs, even if they know profits in subsequent years will suffer.

Managing to cut costs often conflicts with managing to increase throughput. Today, many companies focus on the cost-cutting side of the cloud. The win-win solution to the cloud will both control costs and increase throughput. Goldratt argues that success toward that solution requires focusing first on throughput, because cost will become less and less important as throughput increases.

2.3.2 The production solution

Dr. Goldratt's first career was as a developer of computer software for factory management. He built a very successful business, and his clients were quite satisfied with the software; it gave them much more detailed information about where things were in their factories. He noticed after a while, however, that they were not making any more money using his software. He thought about that and realized that he had to derive the basic principle from a focus on the goal of a for-profit company, that is, to make money now and in the future. The goal corresponds to Dr. Deming's meaning of the aim of a system.

Dr. Goldratt's books, most notably his initial international best seller, *The Goal*, demonstrated how he invented and used TOC to develop the elegant drum-buffer-rope method for controlling production. The drum-buffer-rope method is elegant because it is much simpler than the earlier methods of production management that attempted to control the production system through detailed complexity. The drum-buffer-rope system focuses on the dynamics of the production system.

Figure 2.10 illustrates a production system. Compare it to Figure 2.4; note that Figure 2.10 represents the inner workings of the overall business system depicted by Dr. Deming. Production is a subsystem of the overall business system, just as the circulatory system is a subsystem of the human body.

The "drum" is the processing capability of the constraint. It determines the overall throughput of the production process. Recall that

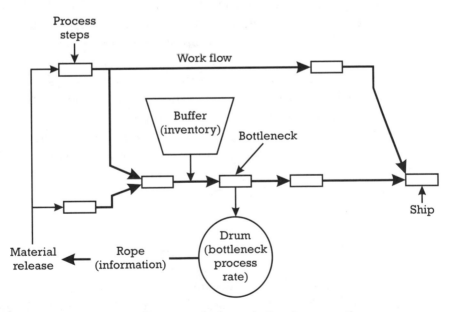

Figure 2.10 Drum-buffer-rope is the solution to operating a production facility using TOC. This solution operates to the global optimum (the system goal) and accounts for the combination of statistical fluctuations and dependent events.

throughput is the difference between sales revenue and raw material cost. To exploit (make maximum use of) the constraint in terms of throughput, you have to release the correct work into the system at the proper time to never starve the constraint and also to not overload it. Overloading the constraint (that is, producing more than it can process) creates excess in process inventory (piles of incomplete work in front of the constraint). The "rope" transmits information from the drum to the release of work, in order to never starve the constraint and to limit the build-up of inventory.

The "buffer" is deliberate placement of in-process inventory to account for statistical fluctuations in the process system. Machines break, go out of alignment, or sometimes need unplanned maintenance. People do not always show up on time and do not work to a constant rate. The buffers account for those fluctuations.

In *The Goal*, Dr. Goldratt used the background of a factory that produces hardware products; however, the general nature of Figures 2.1 and 2.4 works for any kind of system. The output is anything an organization does that it sends outside. Output includes scientific research results, services of any kind, meetings, travel arrangements, reports, legal aid, software products, or any other output of any profit or nonprofit organization. The systems include government. Nonprofit and government systems obviously have a goal (aim) that is different from that for profit business.

Figures 2.4 and 2.10 are static pictures of a production system. The system stays fixed. Inputs flow through the system, converting to outputs. The flow through the system is not uniform. Each step in the processes has some amount of variation, often referred to as statistical fluctuations. Because workstations downstream of other workstations need the parts from the upstream workstations, they are dependent on the upstream workstation. The combination of dependent events and statistical fluctuations is an important issue in managing the overall system, especially at the constraint.

A system designed with capacity for steps upstream of the constraint equal to capacity of the constraint cannot produce at the capacity of the constraint. The reason is that upstream fluctuations add up, leading to periodic starving of the constraint. The constraint can never make up the lost production, because it is the constraint of the system. Therefore, in an optimum system, all upstream workstations must have excess capacity.

Likewise, all workstations downstream of the constraint must have capacity that exceeds the capacity of the constraint. Otherwise, they can never make up any downside fluctuations in their performance relative to the performance of the constraint. Most of the time, they operate at the capacity of the constraint (the drum for the system), but the excess capacity allows them to catch up when necessary. That means all non-bottleneck machines in a production facility should spend some of their time not working.

That reasoning extends to the conclusion that a system operating with each step at optimum efficiency cannot be an efficient system. Most people intuitively believe that operating each part of a system at maximum efficiency causes the system to operate at maximum efficiency. You can see that an optimum system has to feed the bottleneck at its capacity and process the downstream parts at the bottleneck's average processing rate. That means that, on average, every nonbottleneck process must operate at lower efficiency than the bottleneck, in order to have reserve capacity to make up for fluctuations.

This understanding is a major reason that TOC is able to make such immediate impact, once people understand it. Managers design and operate most current systems without the critical understanding of TOC. They work to cut costs everywhere, including the capacity of the constraint. They work to improve efficiency everywhere, including workstations upstream of the constraint that may cause the constraint to work on things that do not translate to short-term throughput. Once they understand the theory, identify the constraint, and improve its throughput, the system throughput increases immediately.

The computer systems that Dr. Goldratt was selling before he invented TOC, as well as all other factory control systems, failed to account for the impact of the system constraint combined with these statistical fluctuations and workstation dependency. Because the actual fluctuations are statistical, they are unpredictable. You can predict only the general behavior over a period of time and many items that flow through the system. Therefore, the schedules produced by the computer systems were outdated and incorrect as soon as they were produced. No wonder the schedule did not cause the system to make more money. No wonder that adding more detail to project plans does not make projects more successful.

In *Critical Chain*, Dr. Goldratt extended the concept of drum-buffer-rope to project planning and performance. It is not a direct extension, because project work on activities moves through time, while in a production facility the parts move through fixed workstations. The same constraint phenomena apply to projects. The combination of statistical fluctuations and dependent events exists in a project. Current computer planning and control methods do not consider those fluctuations.[1] Therefore, many of the same phenomena take place in projects that took place in production before drum-buffer-rope, that is, late delivery, longer and longer delivery times, resources not available when needed, and so on. More detailed planning or more sophisticated computer programs cannot correct those problems because of the structure of the project reality. You do not reduce uncertainty by cutting up tasks. (Remember the fifth-discipline law about elephants.) More detailed plans increase static complexity but do not help deal with dynamic variation due to uncertain estimates.

For a project, the critical chain is the constraint. It is the focus for management of the system. The buffers are time buffers instead of material. (Actually, in production the physical material buffers relate to time also. A pile of a certain size provides a certain time of protection for the machine that works on the pile.) Project buffer management is similar to the production counterpart. Counterparts to the rope are the following:

▶ Release of activities for work based on getting the input from the upstream activity;

▶ Critical chain resource buffers;

▶ The decisions made in buffer management.

Many people are unable to apply TOC understanding to their work. They can see from *The Goal* how to apply it to a physical production system but cannot see it in their system, which may be a service business, research and development, nonprofit organization, or government agency. There is no basis for the distinction; the theory applies to any

---

1. Many computer tools have been developed to analyze the uncertainty in estimates. They are included in some "premium" software, and available as add-ons to other software. In both cases, however, they are usually not used as day-to-day planning and control tools.

system. Goldratt's *It's Not Luck* [15] shows how TOC tools apply to marketing, personal career planning, and personal issues at home.

Experience demonstrates that even in production systems, the constraint usually turns out to be a policy, not the physical bottleneck. *The Goal* [14] demonstrated that relative to financial and sales policies.

Consider a service business that answers telephone calls from customers. A common measure for such services is the number of calls per hour handled by each person. The goal of the system does not relate to the number of calls, but to some effect from answering the calls, for example, satisfied customers or orders. Calls have statistical fluctuations in their length, and they arrive at random times. Suppose you are a customer and want to order many things. Should the operator keep you on the line and thus get marked down for fewer calls per hour? How long will you wait for an operator to answer before you call a competitor?

As the manager of this service, how do you decide when you get more operators? If you have excess operators (to handle longer calls and the variations in when calls arrive), that means your efficiency goes down, even though the throughput for the company may go up far more than the added operating expense. What is the constraint to this system?

Consider another case representative of many internal functions in a company, the human resource function. What is your department goal, and how does it relate to the company goal? How do you measure output to ensure you are contributing to the company goal? Do you know where the company constraint is, and how human resources might influence it? Dr. Goldratt defined several necessary conditions for achieving the goal of a company. One of those is to "satisfy and motivate employees now and in the future," a condition that directly affects the throughput of the company. Human resources clearly affect that necessary condition. Human resources also affect operating expense in several ways, including their own contribution (cost) and the effect they may have on company salaries and benefits through salary and benefit policies and union agreements.

2.3.3 Five focusing steps

Having realized the goal of the system and the fact of a constraint, Dr. Goldratt invented five focusing steps as a process to get the most out of a system, in terms of the system goal. Figure 2.11 summarizes the five steps.

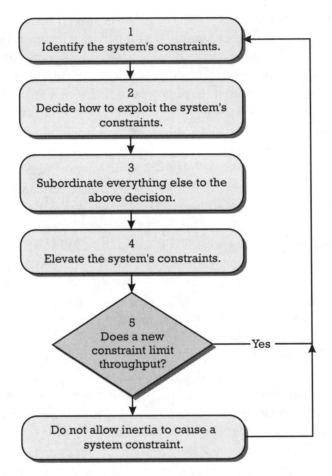

Figure 2.11   The five focusing steps represent the TOC approach to ongoing improvement.

2.3.3.1   Step 1: Identify the system's constraints

To improve the system in terms of the goal, you have to identify what is holding it back. You have to decide what to change. The system's constraint is like the weakest link of a chain: No matter what you do to improve other links in the chain, the chain does not become stronger until you improve the strength of the weakest link. It is evident that you have to find the weakest link before you can improve it.

In a project management system, the weakest link can be anywhere: in the project management process, in company management policies, in any of the supply chains, in work procedures, in the measurement system, or in communication. Because a project does not have physical form until it is well under way, the constraint often is not evident. Systems theory describes why and how symptoms may occur a long time after the actions that caused them. (See the laws of the fifth discipline, in Subsection 2.2.1.1.) You also know that the symptoms may appear somewhere other than the cause, through chains of effect→cause→effect. Therefore, study of why projects have gone wrong may not identify the actual cause of the symptoms.

TOC identifies the constraint of a nonproduction system as a core conflict. Like any constraint, the core conflict is the primary cause of the reasons that the system is not performing better. It is the root cause of one or more undesirable effects in the system. To eliminate undesirable effects, you have to first identify the core conflict.

2.3.3.2   Step 2: Decide how to exploit the system's constraints

Exploiting the system constraint is getting the most out of the weakest link of the chain. There are usually a variety of ways to do that. For example, in a production facility, one way to improve throughput of the production system is to change the way the system puts things through the bottleneck (constraint). It must ensure that policies maximize using the constraint in terms of the goal. For example, ensuring the quality of parts entering the bottleneck prevents the bottleneck from wasting time on defective parts. The schedule ensures that products with the closest delivery date complete first.

For a nonproduction system, you have to decide how to eliminate the core conflict and ensure that you change the necessary parts of the system so the natural effect→cause→effect that results from the changes will achieve the desired effects.

In step 2, you are deciding what to change to.

2.3.3.3   Step 3: Subordinate everything else to the decision made in step 2

Step 3 is the key to focusing your effort. While subordinating, you may find many assumptions that seem to inhibit doing the right thing. For example, in *The Goal*, Alex Rogo discovers many measurement constraints (efficiencies) that would prevent him from doing the right things, if he paid attention to them. The accounting system valued finished-goods inventory as an asset, and it made his financial reports look good to build inventory. In fact, making and storing inventory cost money and can plug up the system's constraint, delaying work that would otherwise go directly to a customer and create income. Likewise, measuring work-stations by efficiencies causes people to build parts for products that are not going to sell immediately, causing cash outlay for parts, and possibly plugging the system constraint, again affecting products that customers want and that would lead to immediate income.

Because project management has been in existence for over 40 years with little change, is it not likely that there are some assumptions, policies, or artificial constraints that do not work well any more? Is it possible that some of the measures used to manage a project actually make it less likely to meet the goal?

Step 3 is the first part of deciding how to cause the change.

2.3.3.4   Step 4: Elevate the system's constraints

Step 4 is the implementing part of how to cause the change. It is often the most difficult part to do, not because of the physical work, but because of the changes it demands in how people look at things. After all, they have been doing things the other way for a long time, without questioning their assumptions. People naturally defend what they have always done.

Sometimes, this defensiveness prevents us from even conceiving of different ways of doing things. It always makes it difficult to implement something new. Chapter 10 discusses how to overcome the five layers of resistance.

2.3.3.5   Step 5: If a constraint is broken in step 4, go back to step 1

As you continue to elevate the current constraint, you always eventually unearth another constraint. It may be lurking a few capacity percentage points above the current constraint, or you may be able to improve the system many tens of percentage points before you uncover the next real constraint. This is not a problem, it just provides a natural strategy to follow in improving a system: Always focus on the current constraint. That is the optimum continuous improvement strategy.

A strong caution follows the five focusing steps: Do not let management's inertia become the system's constraint.

2.3.4 The thinking process

Dr. Goldratt designed the "thinking process" to answer three questions:

▶ What to change?

▶ What to change to?

▶ How to effect the change?

The process steps link, so the output of each step provides the input for the next step. Figure 2.12 illustrates the overall thinking process flow and identifies the primary tools.

Dr. Goldratt developed the tools necessary to apply the thinking process. In addition to their use in the thinking process, the tools (other than the current-reality tree and the future-reality tree) have stand-alone application. This chapter describes the tools, but most of them are not used until near the end of this book, to keep the text accessible to readers who may not be interested in learning more about TOC but would like to improve their projects. The text does make extensive use of the evaporating cloud, the most elegant stand-alone TOC tool. The final chapter demonstrates application of the thinking process to create CCPM. Dettmer [13] provides an effective description and set of procedures to apply the thinking process.

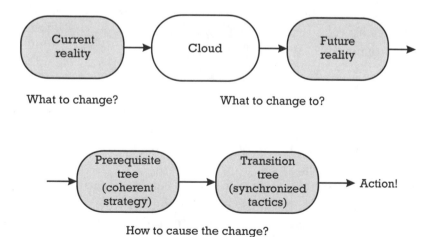

Figure 2.12 Goldratt's thinking process leads from undesired effects, through the core problem, to successful implementation.

Most people find the list of TOC thinking process tools and associated acronyms intimidating at first. Most people require two to three weeks of intensive training and practice to be able to solo with the thinking process and usually several years of applications to become proficient. Noreen, Smith, and Mackey [16] reported that, even after such training, only a limited number of people are able to create significant solutions. (Their book is somewhat dated, and the process, tools, and training have changed significantly since their survey. I am not aware, however, of more recent survey data.)

You do not need to understand the TOC tools to successfully apply CCPM. The reason for describing them at this point is to let you know that CCPM was developed as a robust theory and subjected to extensive critical discussion before it was put to the test.

2.3.4.1 Current-reality tree

The current-reality tree (CRT) is a logical effect→cause→effect model of the existing system that connects a core conflict to a set of undesired effects. Relating all (or most) of the undesired effects of the system to a single core conflict focuses on the leverage point of the system, identifying what to change. Guidelines for scrutiny (in Popper's words, "critical discussion") of the CRT lead to team agreement on the effect→cause→effect relationships that cause the system undesired effects. In other words, it leads to agreement on the right problem. The CRT identifies the policies, measures, and behaviors that contribute to current reality. You read the CRT from the bottom up using IF-THEN logical statements.

2.3.4.2 Evaporating cloud

The evaporating cloud and the guidelines for its communication and use define and aid resolution of conflicts and dilemmas. You can consider it a fixed-format horizontal tree of necessity. You read the evaporating cloud from left to right, using "To have X, you must first have Y" necessity logic.

Dr. Goldratt's evaporating cloud is a good tool to unearth the underlying beliefs or mindsets that cause conflicts and dilemmas.

Figure 2.13 presents a general version of the evaporating cloud in terms of beliefs and actions. (I have come to understand this is as the most basic representation of the evaporating cloud.) The cloud describes two views of reality or two arguments (in the logical argument sense).

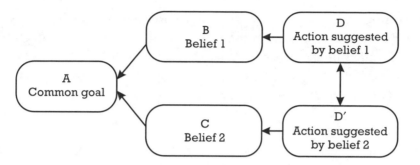

Figure 2.13   Goldratt's evaporating cloud provides a tool to resolve conflicts and dilemmas.

Consider D and D′ as two conflicting propositions about how to achieve the goal. One argument is this: "To have A, I must have B. To have B, I must have D′." The other argument is: "To have A, I must have C. To have C, I must have D′." Thus, even with a common goal, there are two logical ways to get there. The beliefs may be compatible with each other, or they may not. The actions are not compatible. If they were, there would not be a conflict.

The process to resolve the evaporating cloud is as important as the construct. Usually one side constructs the cloud, with a pretty clear view of what the alternative actions are (e.g., D, D′). The constructor side can usually come up with a belief that connects their proposition. They can only guess at the other side's belief. (As noted, neither side may really understand its underlying belief.) The constructor side presents the cloud to the other side, reading the other side first. When reading the other side, the constructor side makes clear that it only guessed at C and accepts any revision proposed by the other side. The constructor side than reads its side, noting, "No wonder we have a disagreement." They then suggest, "Let's search for solutions that will give us A, B, and C and not worry about D and D′. This is a win-win solution. Let's try to identify some assumptions that underlie the arrows in this diagram and see if we can come up with a way to invalidate one or more of those assumptions and get to our win-win solution."

2.3.4.3   Future-reality tree

The future-reality tree (FRT) defines the system you want to change to. The FRT system converts all the undesired effects of current reality into their counterpart desired effects. It identifies the changes that you have to make in current reality to cause the desired effects and provides the effect→cause→effect logic from those changes to the desired effects. The FRT identifies the feedback necessary to maintain the future reality after the actions made to create the injections are no longer active. You read the FRT from the bottom up using IF-THEN logical statements.

2.3.4.4   Negative branch

The negative branch (NBR) aids diagnosis and resolution of single undesired effects. It is a tool to identify and eliminate or correct potential unintended consequences from the changes you make in the system. It is a little tree (thus a branch) connecting some known effect to the undesired effect. Guidelines for scrutiny and buy-in by affected people are the same as for the CRT and the FRT. When used in the thinking process, the NBR starts by assuming successful application of one of the injections made to create the FRT. You read the NBR from the bottom up, using IF-THEN logical statements.

2.3.4.5   Prerequisite tree

The prerequisite tree (PRT) provides a synchronized logical plan to achieve a team objective. It creates team buy-in to the intermediate objectives necessary to reach a higher level objective, such as an injection on the FRT. It makes use of people's natural ability to identify obstacles to achieving objectives and creates a logical sequenced plan to overcome all the obstacles. You read the PRT from the top down, using "To have X, we must first have Y" necessity logic.

2.3.4.6   Transition tree

The transition tree (TRT) provides clear instructions for actions to achieve the objectives specified on a PRT or any objective. It is a logical way to write an effective procedure. The TRT specifies the actions, the reason the action is needed, the result expected from the action, the logic for expecting the action to create the desired result, and the logic for the sequence of the actions. You read the TRT from the bottom up, using IF-THEN logical statements.

2.3.5   Resistance to change

Dr. Goldratt developed a model he called the six layers of resistance to describe the personal aspects of resistance to change. Preceding comments notwithstanding, this is a powerful model when considered in the context of the thinking process, or system analysis, that precedes deciding how to cause the change. This model supplements—not replaces—the system analysis. Dr. Goldratt's six layers of resistance are defined somewhat like the following (he always changes them a little bit):

1. "Not my problem!" (When confronted with a problem, people frequently tend to blame others for the problem and at least disclaim responsibility or accountability to correct the problem.)

2. Not agreeing with the direction of the solution. (People tend to want to stay within existing patterns. If the thing did not work last time, it must be because we did not use it right. Let's try harder next time.)

3. Not agreeing with the specifics of the solution.

4. "Yes, but...." (It will cause some unintended negative consequence, e.g., management or the customer will take away our buffers.)

5. "It'll never work here." (They believe obstacles to implementation are unique to their environment.)

6. Unspecified fear that prevents moving ahead. (Paradigm lock.)

Dr. Goldratt stated that people usually traverse the model in the sequence listed. If they get partway along and feel stuck or lose understanding, they tend to drop all the way back to layer 1, rather than just remaining stuck on the higher level layer of resistance.

2.4 Summary

This chapter showed how thinking from three related management disciplines combines to improve the generic system for project management. There is considerable overlap between these disciplines and little disagreement on fundamental values and principles. I hope you agree from this chapter that:

▶ The PMBOK™ Guide describes a comprehensive project system (current theory).

▶ The principles and practices of TQM and TOC provide tools to improve the theory.

▶ TQM and TOC both operate with Dr. W. Edwards Deming's points of profound knowledge: appreciation for a system, understanding of variation, a theory of knowledge, and understanding of psychology.

▶ The PMBOK™ Guide and supporting literature do not differentiate between special cause variation and common cause variation. Given the topological similarity of the project system to the production system (Chapter 1) and the TOC solution that vastly improved production systems, a similar TOC solution to the project system may remove many of the undesired effects.

▶ TOC provides a logical process to improve a system, answering the questions "What to change?" "What to change to?" and "How to cause the change?"

▶ The TOC five focusing steps provide the steps to implement the improvement process: Identify the constraint, exploit the constraint, subordinate everything else to the constraint, elevate the constraint, and if the constraint is broken, go back to step 1.

▶ Improvement to the project system must first identify the system constraint (core conflict) leading to the undesired effects of the present project system (or current theory). The core conflict will identify what to change.

▶ The TOC thinking process leads to the new system design or what to change to.

▶ The TOC six layers of resistance are useful for defining the final phase of project system improvement, how to cause the change.

The problem definition in Chapter 1 and the theory background in this chapter set the stage to develop an improved theory for project planning and control. TOC provides one tool set, and it provides a strategy to apply the tools of TQM for that purpose.

References

[1]   Hendricks, K. B., and V. R. Singhal, "Don't Count TQM Out—Evidence Shows Implementation Pays Off in a Big Way," *Quality Progress*, April 1999.

[2]   Deming, W. E., *Out of the Crisis*, Cambridge: MIT Press, 1982.

[3]   Deming, W. E., *The New Economics*, Cambridge: MIT Press, 1993.

[4]   Senge, P., *The Fifth Discipline*, New York: Doubleday, 1990.

[5]   Brooks, F. P., *The Mythical Man Month: Essays on Software Engineering*, Reading, MA: Addison-Wesley, 1995.

[6]   Hardin, G., *Filters Against Folly*, New York: Viking, 1985.

[7]   Popper, K. R., *Objective Knowledge, An Evolutionary Approach*, Oxford: Clarendon Press, 1979.

[8]   Shewhart, W. A., *Statistical Method*, New York: Dover, 1986.

[9]   Skinner, B. F., *Science and Human Behavior*, London: The Free Press, Collier Macmillan, 1953.

[10] Kohn, A., *Punished by Rewards*, Boston: Houghton Mifflin, 1993.

[11] Herzberg, F., *Work and the Nature of Man*, Cleveland: World Publishing, 1966.

[12] Goldratt, E. M., *What Is This Thing Called the Theory of Constraints, and How Should It Be Implemented?*, Croton-on-Hudson, NY: North River Press, 1990.

[13] Dettmer, H. W., *Eliyahu M. Goldratt's The Theory of Constraints, A Systems Approach to Continuous Improvement*, Milwaukee, WI: ASQC Quality Press, 1997.

[14] Goldratt, E. M., *The Goal*, Great Barrington, MA: North River Press, 1984.

[15] Goldratt, E. M., *It's Not Luck*, Great Barrington, MA: North River Press, 1994.

[16] Noreen, E., D. Smith, and J. T. Mackey, *The Theory of Constraints and Its Implications for Management Accounting*, Great Barrington, MA: North River Press, 1995.

Contents

The direction of the solution

3.1   Deciding what to change

The most important decision you make when you go about improving anything is what to change. Everything else follows from that decision. If you decide to change something that is not the constraint of the system, you most likely will not affect the system. You could make it worse, by making a new constraint more restrictive than the old constraint. But you can never make the system better by improving a nonconstraint.

Throughout my career I have witnessed dozens of organization structure changes, all attempting to improve the performance of the organization. None of them ever did. I have also witnessed several attempts to improve project management through improved software, more training, or more procedures that failed to achieve significant performance improvement. In each case, the physical change was accomplished—boxes on the organizational chart, people trained, software purchased (and even used), books of procedures—but project performance remained about the same. (Of course, there was also a lot of changing out the managers.) TOC taught me that that can only mean that the solutions did not attack the system constraint. The one experience I had that did result in significant change included many of the features of the critical chain solution. In retrospect, that change would have been much more successful if it had had the full theory and process of CCPM.

3.1.1 Defining the project management system

The goal or aim of the project system is to deliver project results that satisfy all project stakeholders. That requires delivering the promised scope on or before the promised delivery date and at or under the estimated cost. Figure 3.1 shows a black box view of the project system that clarifies the system goal, identifies the system inputs and outputs, and leads to the measures that aid controlling the system to achieve the goal.

3.1.2 Project failure as the undesired effect

The theory of knowledge leads us to define a new problem to improve the project system. Comparing predictions of the current project system (the theory) with reality helps to define the problem. Undesired effects differ from the desired effects necessary to support the goal of successful projects.

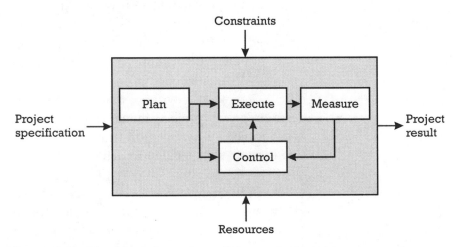

Figure 3.1 The black box view of the project system processes inputs to produce outputs that satisfy the system goal.

UDEs are the things we do not like about the current system. A good way to check them is to use this lead-in: "It really bothers me that ...." Your list of UDEs may not include some of these, and it may include others. Feel free to add or delete as necessary.

- Projects frequently overrun schedule. (If they could put a man on the moon in less than 10 years, how come my projects always overrun?)

- Projects frequently overrun budget. (How hard is it to control cost to what we know we have?)

- Projects frequently have to compromise on scope to deliver on time and within budget. (Wouldn't it be nice just once to complete a project right the first time?)

- Projects have too many changes. (Look, the result we want has not changed, so why am I always signing change approvals?)

- In a multiproject company, projects frequently fight over resources. (You want *who*, for *how* long, to work only on *what*?)

- Project durations get longer and longer. (See first item.)

- Many projects are canceled before they are completed. (A billion here, a billion there, and before long we are talking a lot of money.)

- Project work creates high stress on many participants. (The XXX project brings you greetings.)

TQM and TOC help us to understand that UDEs are a direct result of the project system we are currently using. Even though they are not intended effects, persistence of the UDEs for some time demonstrates that they are robust effects of the system. That means things elsewhere in the system are causing the UDEs. Because the UDEs are observed on all types of projects in all types of businesses in many types of cultures, we can conclude that project type, business type, and culture are not primary factors or influences that cause these results. TOC leads us to suspect that there is some underlying conflict or dilemma that is common to all the environments that exhibit these effects. To decide what to change, you first have to identify this dilemma: the constraint of the current system.

A common practice is to tighten the screws when performance fails to meet expectations. That is, do whatever you were doing, only harder.

(Insanity has been described as doing more of what you have been doing and expecting a different result.)

3.2   Toward a core dilemma

The original TOC thinking process method went directly from the UDEs to create a CRT, a system model of the current reality that was causing the UDEs. The procedure started with any two UDEs and built a logical connection between them. It then added one UDE at a time, filling in the logic until all the UDEs were connected in a system representation of current reality. After a process of what Popper would call critical discussion, the analyst would select an UDE as the core problem and proceed to analyze it as the result of a conflict. That led to an initial change to begin the design of a new system, which no longer created the UDEs and in fact created their opposing desired effects. The process worked, but it was hard and lengthy.

A recent innovation, which Dr. Goldratt indicated was suggested to him by someone else (but whom he did not identify), made the process more direct and seemingly easier to operate by more people. The revised method selects three of the UDEs and analyzes each of them as stemming from a conflict. It then considers the three conflicts together, to define an underlying core conflict. Finally, the revised method uses the core conflict to construct the model of current reality, showing how the core conflict leads to all (or most) of the UDEs in the system. The process concludes with identification of the initial change necessary to begin to revise the system to a future reality free of the UDEs. The following discussion follows this model for three of the project system dilemmas.

3.2.1   Longer and longer project duration

Most people agree that projects seem to take longer and longer. I ask students, "Does everyone know what contingency is?" All participants usually signal that they do indeed understand it. Then I ask someone to define it. A lot of wiggling in place usually follows the question, but eventually someone offers an answer along the lines of "extra time or money to handle the unexpected." I then ask, "Extra compared to what?" More puzzled expressions. I refer to Figure 3.2 as an example of the variation in task performance (which they have seen during a previous estimating

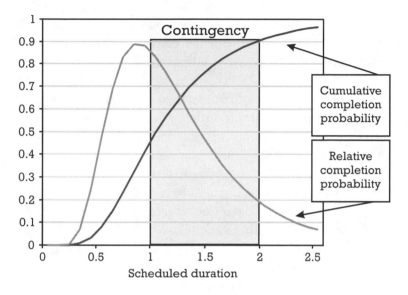

Figure 3.2 Variation in estimates for the time to perform a task helps define contingency.

exercise) and ask, "Isn't it a huge difference if you add contingency to the 50% probable task estimates, as compared to adding it to the 90% probable task estimate?" They all agree and understand that the word *contingency* can have a vast difference in meaning, depending on how you choose to interpret the base. I offer an operational definition: "Contingency is the difference between a 50% probable estimate and a 90% probable estimate." If you do not like that definition, you are welcome to change it. Just be sure that the people you are dealing with use the same meaning.

Everyone wants to have a successful project. One necessary condition to a successful project is to have the project complete on schedule. To have projects complete on schedule, every task on the critical path must complete on schedule. To have every task on the critical path complete on schedule, we must plan each task to include the contingency (as previously defined), because we know that there is uncertainty in task performance. That is the only way to do it with the current CPM. Further, because you find out the critical path only by estimating all the project tasks and connecting the network, you have to include contingency in all the task estimates.

Project managers generally agree that they want people to keep their commitments and deliver on their task delivery date. People generally agree that, in their organizations, people who complete tasks on time are good performers, and people who do not complete on time are considered poor performers. They acknowledge that when project managers ask for input on task times, they want contingency included in the estimates.

Usually, there is also pressure to plan to complete projects as soon as possible. In competitive bid situations, the bidder that can complete sooner usually has an edge. Everyone knows that planning to complete the project sooner tends to reduce project cost, thereby helping make a competitive bid. For those performing R&D projects, the impact of a shorter development may be the difference between the success and the failure of the project. For deadline-driven projects, a shorter plan time usually alleviates the pressure to start now.

For all those reasons, in order to plan a successful project, the project manager must have a shorter critical path for the project. To have a shorter critical path for the project, the project manager must have shorter task estimates that do not include contingency.

Figure 3.3 is the evaporating cloud for that dilemma. Of course, we cannot have both 50% probable task estimates and high-probability task estimates; thus, there is a conflict. In many environments, that conflict plays out by the task estimators proposing high-probability estimates and management, including the project manager, reducing those estimates as a challenge or "stretch" goal. The time cuts usually do not have a method to achieve the time reduction—they are arbitrary. Usually, people know that management still expects them to achieve the low-probability task times. They go into the schedule as fixed dates, and management will request status relative to that date.

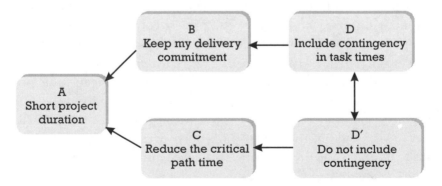

Figure 3.3   Task time conflict.

Task performers tend to accept the challenge. They really have no option. There is considerable pressure to be a team player and to do your part. Subcontractors often have the same pressure—meet the reduced time or we will give the work to someone else. Experienced people justify accepting the situation as a management-dictated version of the chicken game. (Remember those 1950s teenage rebellion movies in which two drivers raced toward a cliff or toward each other to see who would veer off or stop first?) People on a project know that what is happening to them is also happening to every other task on the project. If they agree to the time cut, it is likely that reality will strike some other project task before it gets them, causing management to chicken out and extend the project time. That gives them the time they need to complete their task on time, so they can win in the system. If they were to object to the time cut, they would lose immediately because management would brand them as nonperformers or nonsupporters. They have no choice in the real world of power politics.

3.2.2 Projects frequently overrun schedule

When asked why projects overrun schedules, people usually say the projects start out fine, but somewhere along the way a snag develops that begins to push one or more deliverables later and later. Everyone knows that it takes only one task to be late on the critical path to make the whole project late. As the shift begins to hit the plan, management tries to solve the problem causing the shift, usually diverting resources and making changes in the project plan to cost more and more emphasis on the part of the plan that is slipping. The people working on the snag usually feel a lot of pressure to get their part of the project solved and therefore put in a lot of extra time and feel considerable stress. These are often the resources in most demand in the company, so putting more time on the project in trouble leads them to neglect the other projects they were supposed to be working on, causing other projects to slip as well.

When asked why that happens, people respond with two general types of answers. One type of answer focuses on the specific problem with the specific project that is most recent in memory (often still in trouble). They usually blame it on poor performance by the group responsible for that part of the project. The second type of response is more general, blaming the problem on the tendency of stereotype task performers to underestimate or on management's setting arbitrary completion dates.

How often do people complete activities and pass their work on early? How often do they complete activities for less than the budgeted activity cost? You might find that occurs less frequently than you would expect, if the estimates were truly 90% probable estimates. Even with skewed distributions, tasks should complete early a substantial percentage of the time. Figure 3.4 illustrates typical results for actual times that project tasks, compared to their planned duration. Figure 3.4 shows that most tasks complete exactly on the due date; often, as many as 80% complete on the due date. That is not consistent with the task completion time estimate presented earlier.

Potential causes for little positive variation in activity duration or cost include the following.

▶ People work diligently to milestone dates and do not understand a desire to have the work completed early.

▶ Estimates are much less probable than were believed, leaving little potential for positive variations.

▶ The work expands to fill all available time and budget.

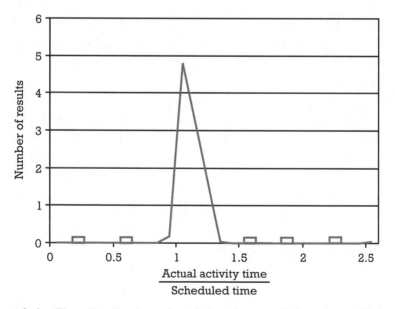

Figure 3.4   The distribution of actual task completion time differs from the estimate distribution and shows a remarkable percentage of completions right on the due date.

▶ Belief that the next activity would not be ready to use it anyway.

▶ In most organizations, there are significant penalties (threatened or real) for completing activities late. But what are the "rewards" for completing activities early? Have you ever seen any of the following:

  ▸ Management gives the product a more detailed review than they would if you completed it near the due date and forces you to make unnecessary changes.

  ▸ The reduced budget for the performing organization leads to higher overhead rates and, in extreme cases, downsizing.

  ▸ Reduced credibility in the performer's activity duration and cost estimates leads to increased pressure to reduce estimates.

Those factors add to the psychological reasons that cause projects to lose much of the potential positive variance. Project managers assign tasks and train people to respond to specific milestone dates. Thus, even if they are done early, they might hold on to the product until the due date. Why not? Management usually does not take any advantage of early completion or reward the task performers if they do deliver early. If the resource performing the work is paid in accordance with the time they spend on the task, they are incentivized to use up all the resources authorized. If your project uses a cost reimbursement contract with them (the usual practice for resources in the company and for certain types of external resources), they may even be incentivized to slow down the work to get overtime pay or more total revenue from the project.

If one resource gets its activity done early, what is the chance that the next critical resource down the line is ready to hop to and start working on its activity? If it is a critical resource, it is in demand and has limits on availability. It does not seem likely that the next resource will be able to work on the activity until the date they had planned for it. Therefore, the positive variance is lost and wait time is introduced. That means that the actual schedule time grows due to activity dependence.

All of that leads to the second conflict, illustrated in Figure 3.5. The upper path refers to the performing resource. To be a successful team member, I must contribute to early completion of the project. To contribute to early completion of the project, I must turn work in early. On the

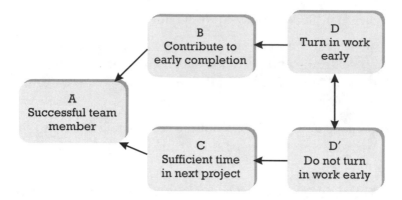

Figure 3.5 The conflict underlying project schedule overruns.

lower branch, to be a successful team member, I must have sufficient time in my task estimates to complete my commitments. To complete my commitments, I must turn in task early. The obvious answer is that I can always do extra checks and improve the quality of my project task result when it looks as if I might finish early. Even if I did finish early and turn it in to my manager to be checked prior to submitting it to the project, the manager (a very busy person) likely would not look at it until it is due anyway.

3.2.2.1 Student syndrome

Did you always study for exams weeks ahead, so you could go to bed early the night before? Did you always write your papers to get them done at least a week before the deadline, to avoid the gap in the library when all the books on the topic are out, and to get to the college computers before everyone else was on them all night?

Well, it is probably not news to you that most people have a tendency to wait until tasks get really urgent before they work on them. That is especially true for busy people in high demand, that is, all the most important people the project manager is counting on to get the critical path work done on time.

Figure 3.6 shows the typical work pattern of many people. They do less than a third of the work on an activity during the first two-thirds of the activity duration. They then do two-thirds of the work during the last third of the activity duration. Where are they more likely to find they have a problem to complete the activity in the remaining time, during the

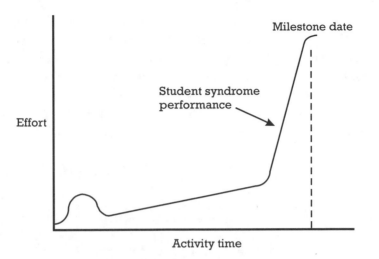

Figure 3.6 People perform most activities, and most people follow the student syndrome performance curve.

first third of the effort or during the last third? If they are working above 100% capacity already to complete two-thirds of the work in one-third of the time, there is no chance to keep to the activity duration by a little extra effort. What is the chance they can recover from an unanticipated problem, like a computer crash?

Student syndrome behavior results in little chance of seeing the positive side of activity duration variation. The effects described above make it unlikely we could take advantage of positive variation, even if we did see it. No wonder projects rarely complete early! The reality is that relative activity duration normally shows a skewed distribution, with a mean well above the average activity time. That is one reason why we often see overruns on activity time but rarely see underruns.

Most project management guidance recommends that project managers use an early start schedule. That means starting all noncritical path activities earlier than is necessary to meet the schedule date. People working on those activities know there is slack in their activity. How does that influence the urgency they feel in working on the activity?

3.2.3 Multitasking

Now assume that the schedule system demands that people working on the project start on activities as soon as possible and report the task start to the project manager. Further assume they split their time during the day evenly to the three activities. When do they complete? If we assume that no time is lost from dropping each activity every day and getting back into it the next day, none of the activities is complete until the third week. Multitasking has increased the activity duration for all three projects to three weeks. They have delayed throughput on the first project for two weeks and on the second for one week. Figures 3.7 and 3.8 illustrate the multitasking conflict and its effects.

While most people will acknowledge that situation, many argue, "It is just not realistic to do otherwise. We have to satisfy multiple needs." They

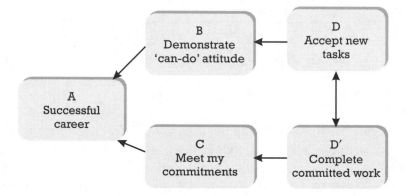

Figure 3.7 The multitasking conflict.

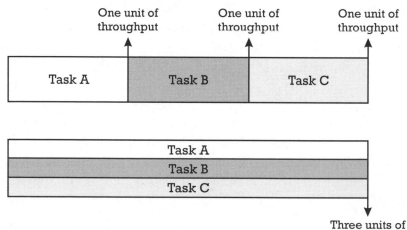

Figure 3.8 Multitasking delays all projects. It also justifies using the longer task times in future plans.

agree with logic that demonstrates that multitasking is a poor (perhaps the worst) way to meet multiple needs. They acknowledge that it deliberately lowers their personal throughput contribution. (And that is still without accounting for the fact that leaving and returning to tasks usually affect the total time necessary to complete the task and often the quality of the product.) Nevertheless, many people find it extremely difficult to change such behavior. Purveyors of time management tools work to resolve this personal conflict at the personal level.

Marris contends that behavior such as multitasking is a social effect of the more powerful using the less powerful to shield them from uncertainty [1]. In other words, management takes advantage of the lower level resources in the organization by creating the pressure to work on its latest idea, leading to multitasking.

3.2.4 Core conflict leading to UDEs

You can combine the three conflicts to obtain the underlying core conflict leading to all three conflicts examined. Because the conflicts derived from the three starting UDEs, resolving the core conflict should have a desirable impact on all three of the UDEs analyzed. Because the project system is a connected system, the core conflict may contribute to the other UDEs as well.

Figure 3.9 illustrates development of the core conflict. The goal of the three conflicts is common: project success. The top path of the cloud illustrates the logic that leads to each individual, and therefore each individual task in the project, to work toward its own success. To have a successful project, each task must perform as planned. For each task to perform as planned, each task performer must do whatever it takes for individual task success.

The lower path illustrates the logic that leads to working toward project success. For the project to succeed, each part of the project must contribute to overall project success. To contribute to overall project success, each task must subordinate to the overall project.

That core conflict is the common conflict Dr. Deming referred to, in which working for each part of the system does not lead to an effective system. It is the conflict identified as a principle in TOC: An optimum system cannot have each part of the system as an optimum. Worse yet, the core conflict sets up a win-lose situation between all the project

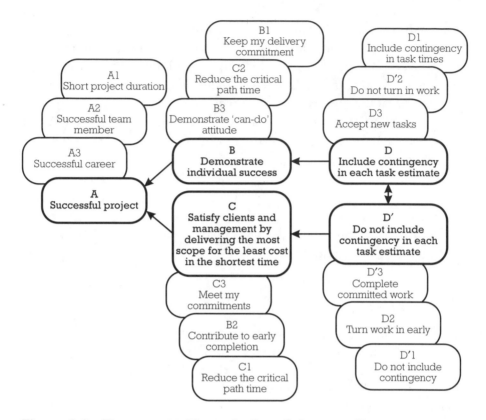

Figure 3.9   The core conflict underlies all three conflicts.

workers and the project management. No wonder that projects are so stressful to all concerned; no wonder so many projects fail.

Figure 3.10 illustrates how the core conflict leads to all the UDEs. It implies that the core conflict is a high-leverage part of the project system. A solution (new theory) that resolves the core conflict differently can influence the whole system in a way that tends to move the UDEs to their desirable counterpart.

The logic illustrated by Figure 3.10 is incomplete. It is only a notional connection between some part of the core conflict and the UDE. At this point, if you accept that the core conflict underlies most of or all the UDEs of the project system, you may be willing to consider the beginning of the solution direction.

How do we know that this is *the* core conflict and not the result of some biased analysis? After all, real experts in project management have

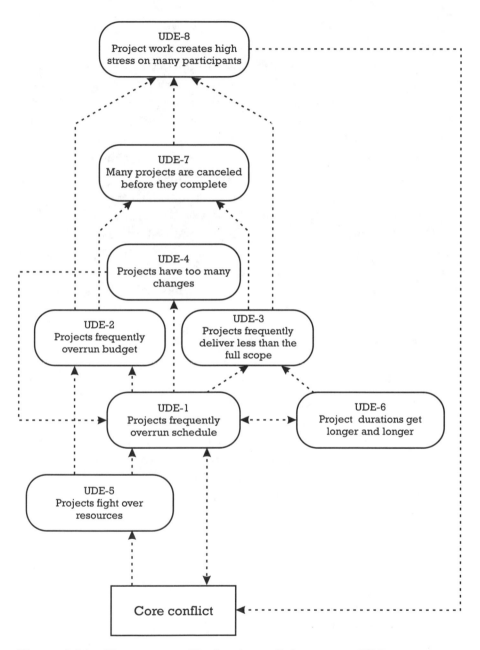

Figure 3.10   The core conflict leads to all the system UDEs.

been working on the problem for years, and this is not the result they have come up with. Usually, they say it is some form of not effectively implementing the project management system roughly defined by the PMBOK^{TM}. Others claim it is all about the project leader.

You can never prove that it is *the* core problem. In a dynamic system (chicken→egg→chicken), there usually are only circular correlated effects. You need an effective starting point to change such a system. But because the system is completely interconnected, starting anywhere will affect the whole system. To carry the example further, in the past breeders affected the chicken and egg system by selecting the chickens they would breed. Now, they can directly affect the genes in the egg. The thinking process is looking for a core conflict that seems to have the most—or at least a lot of—impact on the UDEs of the system.

The core conflict is a hypothesis about the system. It is a theory. To find out if it works better than other theories, you have to subject it to critical review and, ultimately, test. The test must predict something happening with this theory that does not happen with the old theory.

3.3 Toward desired effects

3.3.1 Resolving the core conflict

Resolving the core conflict requires identifying one or more assumptions that can be made invalid by changing the system. Assumptions underlie each arrow of the core conflict. The critical chain method arises from attacking the assumption that adding contingency to each task is the only way to manage uncertainty.

Dr. Goldratt was uniquely positioned to develop the critical chain solution for projects. The critical chain solution comes from recognizing that the variation in task performance and dependent events is at the root of the behavior of the current system. He had tremendous success in applying the solution for production management he described in *The Goal* [2]. He knew that in most cases the uncertainty in project duration estimates is much larger than the variation in production processes. He also knew that in many cases the task dependencies in projects were equal to or greater than the dependencies that exist in production. It is natural that he would look at projects from this perspective to find the assumption to attack.

In *The Goal*, Dr. Goldratt describes the impact of variation and dependent events in the saga of Herbie. He used the scenario of a troop of boy scouts on a hike through the woods. The trail is narrow, so no scout can pass the scout in front of him. As they hike, the line grows longer and longer. Alex Rogo, our hero in *The Goal* and the troop leader for this weekend, realizes what is happening. The speed of the scouts is not the same. There are statistical fluctuations in how fast the scouts walk. Each is dependent on the scout in front of him and the one behind him. These fluctuations cause the length of the line to grow continuously. Herbie turns out to be the slowest boy scout, the constraint. The gaps in the line compare to inventory in a manufacturing plant.

For a project, the gaps in the line of boy scouts compare to time. If the next resource is not ready to start when a predecessor activity completes early, the project loses time. We lose the positive variances in statistical fluctuations. That is like a faster boy behind a slower one; he can catch up but not pass. The line grows in length. This is worse in a project than in a manufacturing case. In manufacturing, the inventory is used eventually. In a project, the time is lost forever. There is no conservation of time.

The direction of the solution Goldratt proposed follows from his TOC production solution. The first step is to identify the constraint of the project system. His focus on throughput led him to focus on the time it takes to complete the project. The longest path through the project is the evident constraint. At first look, this is the critical path.

How then to exploit the critical path? Dr. Goldratt, who holds a Ph.D. in physics, knows statistics and knows a lot about the cloudy behavior of much of reality. He knows that the only way to take advantage of statistical knowledge is through dealing with numbers of events. Deming and Shewhart before him had pointed out that science cannot make predictions about a single instance of a statistical event. That leads to a simple (in retrospect) insight: Concentrate the uncertainty for many of the tasks of the project at the end of the project in a buffer. The buffer has a direct counterpart in his production solution, where buffers of in-process inventory are strategically placed in front of machines to prevent them from running out of work.

Concentrating contingency in the buffer brings along two significant bonuses. The first bonus is a shorter plan. It is a mathematical fact the variances of the sum of samples from a series of independent distributions add. The variance is the square of the standard deviation. The standard deviation is proportional to the amount of variation in a single task. In other words, the uncertainty in the sum of tasks is the square root of the sum of the squares of the individual variation. While attempting to protect the completion date of each task in a project, each task had to include its own allowance for uncertainty. Those allowances add up down the path. When we take the allowances out of each task and put them at the end of the path, they add as the square root of the sum of the squares, a much smaller total amount. Figure 3.11 illustrates how that works for a very simple case. The reason is evident. Some of the tasks should overrun, some should underrun. The distribution of the sum need not be as large as the sum of the individual variations because some will cancel out.

A second statistical fact comes into play with this strategy. The central limit theorem of statistics states that the distribution of samples from a variety of independent distributions tends toward a normal distribution. A normal distribution is a symmetrical distribution. It does not have the long tail to the right that many individual task distributions may have. That means concentrating contingency at the end of a path reduces the likelihood that it will be overrun by a large amount.

A key part of the direction of the solution Goldratt proposed, then, is to use average task completion times in the plan and to add an aggregated buffer at the end of the plan for overall project contingency.

3.3.2 The resource constraint

Eliminating contingency from the individual task estimates and controlling it at the project (path or chain) level appears to directly resolve

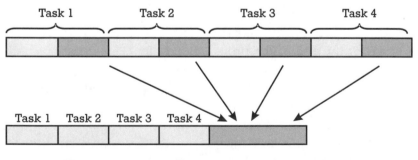

We need less time this way, because of our profound
knowledge understanding of variation!

Figure 3.11 Concentrating contingency at the end of the path requires less total project time.

the first two conflicts that led to the core conflict. It does not resolve the multitasking problem. To avoid multitasking within a single project, you have to eliminate demands on the resources to multitask. To avoid multitasking in an environment with multiple projects, you have to either eliminate the demand or give the resources a way to handle the multiple demands without multitasking and without negative reinforcement.

In the past, I never questioned the proposition that an acceptable way to remove resource contention is to first identify the critical path and then level resources. Resource leveling moves around tasks in the plan to match the resource demand for each resource (e.g., engineers) to the supply. My literature search did not reveal the basis for the proposition that you can first find the critical path and then level. I suspect—but have no proof—that may be a result of technological evolution. It is possible to calculate a project manually to find the critical path. There is no simple algorithm to create an optimum resource-leveled critical path. Thus, it is difficult to resource level even a modestly complex project plan manually. The relatively expensive and slow computers that existed at the time of the growth of CPM and PERT did not lend themselves to doing a lot of calculation. The idea that you could use the computer to calculate the critical path, lay out the network, and then deal with the potential resource constraint seems logical enough. It may even have been that, for projects using CPM and PERT, resources were less often a constraint. They could find the critical path and then determine and satisfy the resource demand.

Current project management software operates by starting with the activity structure (critical path) and only then considers the limited resources available for the project. Project management software identifies the critical path by linking the project activities in a logical way and then measuring the longest time through the network of activities, assuming no resource constraints. The project manager inputs resource availability. The software then allocates the resources through various schemes but usually first to the critical path and then to the paths that are nearest to the critical path in time duration (minimum slack activities first). People who have studied resource allocation know that this does not always give the optimum schedule. People have proposed various heuristics, and some programs provide a large number of selections. The only way to find the optimum among those options is trial and error. Critical chain scheduling resolves the dilemma.

Consider, for example, the miniproject illustrated in Figure 3.12 and determine the critical path through the project. The project has three paths, all of which have two activities with different time duration, as shown. What would we come up with as the critical path? The lower path, that is, activity C1 followed by activity C2, is longest. The project should complete in 65 days. We confirmed this simple calculation with Microsoft Project.

Now, let us move from the world of unlimited resources to the world of reality. Figure 3.13 shows the resources needed and for each activity. There is only one clear resource and one crosshatched resource for the project. When people work on the activity, they have to work full time. What will our critical path schedule now say for the earliest completion time?

If you came up with 160 days, you are applying resource leveling the way most computer programs do. We tested with Microsoft Project and got the same answer. You first apply a conflicting resource to the critical path. You next apply it to the activity with the least slack, in this case, B1.

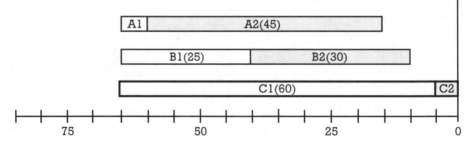

Figure 3.12 Example project shows resource conflict.

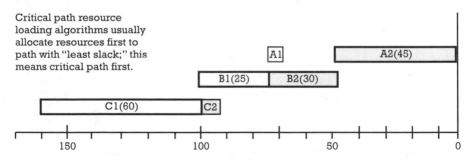

Figure 3.13 Resource loading the critical path for the project yields a schedule of 160 days.

Oh, did you check the crosshatched resource also? Does your solution look like Figure 3.13?

Now, see if you can outsmart the computer. Look at Figure 3.14 and see if you can find a better way to sequence the activities that will reduce the overall project time. You are now considering *both* the activity logic and the resource constraint.

Considering the resource logic with the task structure reduced the critical path time to 95 days. Microsoft Project allowed us to move the tasks to accomplish that and confirmed that it results in resource leveling. Allocating resources using the common critical path method can lead to excessively long schedules.

Thus, Goldratt's second key insight is to define the project critical chain instead of the critical path. The critical chain includes both the task and the resource constraint.

3.4 Solution feasibility (evidence)

Using the scientific method as the theory of knowledge leads to selecting the preferred theory through critical discussion and test. Comparing critical chain to critical path, we see more content in the critical chain theory because it:

 ▶ Provides an explicit method to manage common cause uncertainty;

 ▶ Explicitly resolves the resource constraint.

Popper noted that a new theory should contain and explain the old [3]. With unlimited resources, the critical chain is the same path as the

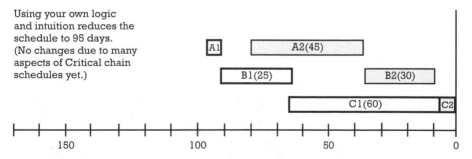

Figure 3.14 Using logic to work out the critical chain reduces the project lead time to 95 days.

critical path. With a resource constraint, the critical chain is an acceptable solution to the resource-leveled critical path. Thus, critical chain contains the critical path solution.

Popper suggested that the primary method of testing a new theory be through critical discussion. Such discussion checks the new theory against the old, looking for logical deductive reasoning and evidence supporting the suppositions (assumptions) made in the new theory and the old theory. Summarizing the reality of the scientific method, Popper states [3]:

1. Induction, that is, inference based on many observations, is a myth. It is neither a psychological fact, nor a fact of ordinary life, nor one of scientific procedure.

2. The actual procedure of science is to operate with conjectures: to jump to conclusions, often after one single observation.

3. Repeated observations and experiments function in science as tests of our conjectures or hypotheses, that is, as attempted refutation.

This chapter developed the reasoning behind the way Goldratt defined the problem with the current theory. It does not explain the jump to his proposed direction for the solution: improved management of uncertainty. It is unlikely that others without his knowledge and experience could have made the same jump. The original PERT method and subsequent work with project simulations are evidence that others were aware of the uncertainty problem. The current knowledge base lumps the uncertainty in predicting each project task in the area of risk management, adding evidence that people understand the need to deal with it. However, none of the current solutions make uncertainty management part of the basic project system in the manner of critical chain.

Critical chain explains the reasons for schedule overrun through the reality of statistical fluctuations (uncertainty or variation), dependent events, and human behavior. The CPM theory does not address that reality; it uses deterministic durations and start and stop dates for each activity in the schedule. Combining that technical assumption with human behavior leads to schedule overrun. Schedule overrun leads to cost overrun and reduces the delivered scope. Perhaps most important, the new theory explains how the CPM theory, through the win-lose approach to task scheduling, causes much of the psychological harm in project systems.

The resource constraint is every bit as real as the task predecessor constraint. It is a necessary condition to perform the task. The critical path method assumes that an acceptable solution to the resource constraint is to first find the unrestrained critical path and then assess the impact of the resource constraint. Put another way, determination of the critical path assumes that resources are not the constraint. Or it assumes infinite resources. We could find no references describing the reasoning behind that assumption. Goldratt found it easy to notice this implicit assumption because it was made as an explicit assumption in the production system models he had worked with [4].

This chapter demonstrated by example how the CPM to resolve the potential resource constraint could lead to poor plans. The larger concern from a TOC perspective is that the method therefore never identifies the real constraint to the project. It is a simple and logical step to define the critical chain as the combination of the two potential constraints on the longest path to complete the project.

Chapter 1 presented selected successful evidence that the critical chain method creates the desired effects. (It is selected in the sense that it is not an exhaustive listing; that does not mean we selected only the positive results!) By this time, hundreds of projects of different types, in different businesses, and in cultures around the world have successfully applied critical chain. I am not aware of any cases in which project failure has been attributed to critical chain. I am aware of several cases in which implementation failed to achieve the changes necessary for critical chain, and the project system continued to operate the old way. Chapter 9 addresses that in detail.

3.5 Determining what to change to

Resolving the core conflict provides a necessary change to the system to begin to move toward the desired effects. The desired effects for the project system, derived from the current-reality UDEs, are these:

- ▶ Projects always complete on or before the scheduled completion date;

- ▶ Projects complete within or for less than the budget;

- ▶ Projects always deliver the full scope;

- ▶ Projects have few changes;

- ▶ Projects have needed resources without internal fights;

- ▶ Project durations get shorter and shorter;

- ▶ All projects complete;

- ▶ Project work creates win-win solutions for all stakeholders.

The changes to resolve the core conflict provide a method to manage uncertainty and acknowledge the reality of the resource constraint that affects many projects. The changes by themselves are not sufficient to create all those desired effects. While the solution to the core conflict explicitly considers the project system and addresses variation, it does not address all the psychological elements that affect project performance. Subsequent chapters provide the complete solution leading to all these desired effects, and Chapter 11 provides the complete logical development of the full solution.

3.6   Summary

This chapter identified the core conflict for project management system improvement as the way the project system manages uncertainty. Specific points made in developing that theory are:

- ▶ Undesired effects define the problem with the present project management theory.

- ▶ The core conflict for the current project management system is the conflict between protecting each task duration in the project versus protecting the entire project.

- ▶ The project system solution (new theory) must overcome the core conflict of the existing project system by attacking a key assumption in present systems: how to manage uncertainty.

▶ A method to manage uncertainty using ideas similar to those that succeeded in improving production performance may provide a direction for the solution.

▶ The direction of the solution should be to manage uncertainty by concentrating contingency into buffers at the end of chains of tasks.

▶ Because the resource constraint is equal to the task logic constraint, the longest path through the project, the critical chain, should include both.

▶ A growing body of empirical evidence demonstrates the feasibility of the critical chain project management method for all types of projects.

Note that managing uncertainty is not the same as knowing about uncertainty or analyzing uncertainty. People knew about uncertainty long before projects began. There are many methods to analyze uncertainty. Both knowledge and analysis are necessary to manage, but they are not sufficient. You are managing only when you take actions that drive the system to the goal. Chapter 4 derives the full system to do that.

References

[1]   Marris, P., *The Politics of Uncertainty*, London: Routledge, 1996.

[2]   Goldratt, E. M., *The Goal*, Great Barrington, MA: North River Press, 1984.

[3]   Popper, K. R., *Objective Knowledge, An Evolutionary Approach*, Oxford: Clarendon Press, 1997, p.144.

[4]   Goldratt, E. M., *Theory of Constraints*, Great Barrington, MA: North River Press, 1994.

CHAPTER

4

Contents

The complete single-project solution

This chapter describes the process to develop the single-project management system to satisfy the system requirements identified in the previous chapters and further defined herein. Although presented as a forward moving process from requirements to design, the actual process, as in nearly all designs, was iterative. That is, various design solutions were proposed and tested against the requirements until a suitable working system resulted.

4.1 From system requirements to system design

4.1.1 Requirements matrix

Table 4.1 illustrates the requirements for an effective project management system, following the method of Joseph Juran [1]. Table 4.1 presents the requirements in a hierarchy,

Table 4.1(a)
Overall Requirements for a Project System to Convert the Input of a
Project Result Specification and Produce an Output of a Completed
Project Result

Primary Requirement	Secondary Requirement	Tertiary Requirements
Define the project system	Project system goal	Define the project system goal to complete projects that make money for the company, now and in the future (for profit companies)
—	Project system boundary	Define the project system boundary starting with customer needs and ending with a satisfied customer
—	Account for understanding of variation	1. Account for common cause variation in project processes 2. Provide a means to separate and deal with special cause variation
—	Use TOC to design the system	1. Identify the project constraint 2. Exploit the project system constraint 3. Subordinate everything else
—	Include knowledge of psychology in the system design	1. Align project system needs with individual psychological needs 2. Align individual rewards with project system needs
—	Enable continuous improvement of the project system (a theory of knowledge)	1. Define and standardize processes 2. Measure process performance 3. Assess process performance 4. Improve processes

starting with the top-level necessary conditions for project performance. These conditions include the three technical requirements for the project (scope, cost, and schedule) and the requirement for stakeholder satisfaction. The table is segmented to make it easier to read and understand: the first segment provides general system requirements, the second segment provides project necessary condition requirements, and the third section provides stakeholder requirements. Project stakeholders always include at least the project customer and the project team and may include many others, for example, subcontractors, stockholders, regulators, neighbors, or government. The second and third columns of the table illustrate the second- and third-level requirements derived from the top-level requirements. Requirements at the lower levels may vary for different types of projects; these are general requirements.

Table 4.1(b)

Necessary Conditions for a Project System to Convert the Input of a Project Result Specification and Produce an Output of a Completed Project Result

Primary Requirement	Secondary Requirement	Tertiary Requirements
Deliver the project result to the specification (scope)	Deliver all of the specified features	1. Satisfy all the physical requirements for the specified features 2. Satisfy all the functional requirements for the specified features 3. Satisfy all the operational requirements for the specified features
—	Satisfy all the feature quality requirements	Satisfy all the feature quality requirements
Deliver the project result on time (schedule)	Deliver the project result on time (schedule)	Complete the project on the quoted completion date
—	—	Complete intermediate milestones on the quoted completion dates
Deliver the project result for the estimated cost	Total cost	1. Complete the entire project for the quoted maximum cost 2. Do not spend more than specified maximums on subcategories of the total cost
—	Satisfy project cash flow requirements	Do not exceed project estimated cash flow requirements

It is unlikely that you would generate an identical list of project requirements. The lists in Table 4.1 include elements of the PMBOK™ elements from my own experience and elements specifically derived from the solution we are about to present. This feedback of the solution to the requirements is part of reality. Only by defining and critically assessing a proposed solution do we really understand the problem. In particular, before Goldratt and considering the basis of critical chain, I would not have included accounting for common cause variation among project requirements. Instead, I would have combined it with project risk management, as PMBOK™ does.

The table of project requirements is not—and never can be—complete. It is a conjecture, a basis for criticism and improvement. For example, I am not satisfied that the requirements completely embrace profound knowledge, especially a knowledge of psychology. I present it

Table 4.1(c)
Stakeholder Requirements for a Project System to Convert the Input of
a Project Result Specification and Produce an Output of a Completed
Project Result

Primary Requirement	Secondary Requirement	Tertiary Requirements
Satisfy unique individual project stakeholder needs, in addition to those listed in (a) and (b)	Project client	1. Solicit and specify all requirements necessary to deliver a satisfactory final product 2. Provide evidence of meeting the project specifications 3. Provide information during the project to enable decisions that may affect the balance of the project 4. Respond to requests for changes
—	Project team	1. Clear scope definition 2. Clear responsibility and authority assignment 3. Project plan specifying who has to do what by when 4. Feedback to control performance to plan 5. Method to control interfaces with other team members 6. Method to raise and resolve issues during project performance 7. Change control process

as a good enough set of requirements to bind together CCPM and start us on a new path of project system improvement to address the difficulties raised in Chapter 1.

4.1.2 Summary of single-project critical chain

Figure 4.1 illustrates the key features of the single-project critical chain solution that satisfy the functional requirements for the project system. The illustrated features highlight the differences between CCPM and critical path planning and management. Those essential features are:

▶ Defining the critical chain as the longest path through the project considering both the task logic and the resource constraint;

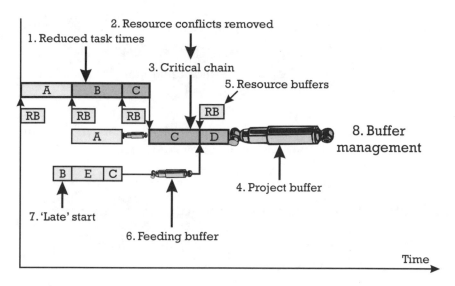

Figure 4.1 Key features of the critical chain solution deliver performance to the project system requirements.

- Removing resource contention from the project plan before selecting the critical chain;

- Developing the plan with 50-50 task estimates, aggregating uncertainty into the buffers at the end of task chains (buffer illustrated as a shock absorber);

- Protecting merging paths with feeding buffers (while continuing the elimination of resource conflicts);

- Adding resource buffers to ensure that critical chain resources are available when needed;

- Using the project and feeding buffers as measures to control project performance.

The next section describes each of those features in greater detail. Four essential behavior changes are required to use CCPM effectively:

- Management must encourage the use of 50-50 task times by not pressuring people to perform to 50-50 estimated durations.

- Management must enable people to focus on one task at a time.

▶ Resources must focus on one task at a time and pass on the results as soon as the task is complete (roadrunner behavior).

▶ Everyone must use the plan and the buffer reports to decide what to work on next.

Although those behavior changes are simple, they are not necessarily easy to implement. (Chapter 9 covers implementation.)

4.2   Developing the critical chain solution

This section describes the single-project critical chain features in terms of the TOC focusing steps (described in Section 2.3.3).

4.2.1   Identifying the project constraint

Defining the constraint of a project in terms of the schedule derives from the impact that schedule has on project cost and project scope. Independent variables that influence a project result include the demanded scope, the project system definition, and the resources available to work on the project. The project system outputs are dependent variables (delivered scope, cost, and schedule). As schedule increases with fixed deliverable scope, cost usually increases. As scope increases with fixed cost (or resources), schedule tends to increase. As scope increases with fixed schedule, cost tends to increase. Therefore, it is appropriate to focus first on delivering the project on time.

The evident constraint of a project is the chain of tasks that takes the longest to complete. To perform any task on a project, two things are necessary: the task input from a predecessor and the resource to perform the task. (The predecessor may simply be a start authorization for the first task in a chain of project tasks.) The definition of the critical path does not explicitly address the potential resource constraint. Section 3.3.2 described why the critical chain is the single-project constraint. The critical chain is simply "the sequence of dependent events that prevents the project from completing in a shorter interval. Resource dependencies determine the critical chain as much as do task dependencies."

Figure 4.2 illustrates a typical critical path project schedule. The letters represent unique resources. For this illustration, assume there is only one

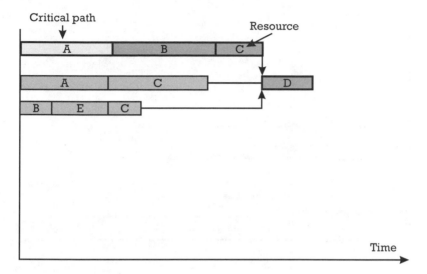

Figure 4.2 The critical path does not account for the resource constraint.

resource corresponding to each letter, that is, only one A resource, only one B resource, and so on. Evidently, we would fail to meet the schedule on the project, because each resource can do only one task at a time.

Figure 4.3 moves tasks to eliminate the overlap of resource demands. In a manner similar to many computer algorithms for resource leveling, we first give the resource to the path with least float; usually with the initial critical path. Note that when we are done, all paths have float, so there is no critical path defined as the path with zero float. (Computer software packages treat that result differently. Some keep the initial critical path definition. Some define only the last task as critical. I have no idea what they do about the critical path as the project progresses and the critical path is supposed to change.)

More importantly, the initial critical path is not the constraint to completing the project. Because the resource constraint is often a significant project constraint, the TOC method of project planning always considers it (Figure 4.4).

If your organization does not have resource constraints (or has infinite resources), the critical chain will be the same initial task path as the critical path. That is an important fact in verifying the integrity of the critical chain method; it contains the critical path method as a special case, at least in regard to defining the critical chain.

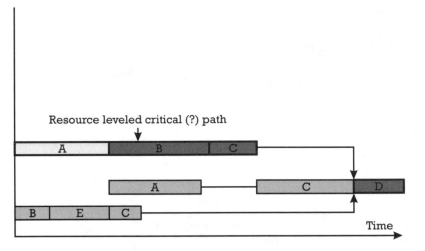

Figure 4.3 Removing resource conflicts usually creates gaps in the critical path.

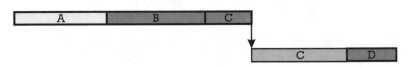

Figure 4.4 The critical chain includes both the resource and task logic constraint to completing the project on time or sooner.

The PMBOK™ Guide definition of critical path states that the critical path may change during the performance of the project. That occurs when project tasks experience common cause variation and redefine the longest zero float path to complete the project. Based on our knowledge of variation, that means we should expect the apparent critical path to change frequently. Dr. Deming noted that one of the more serious mistakes managers can make is to treat common cause variation as if it were special cause variation. That PMBOK™ Guide definition of critical path (and implementation in many project management systems) institutionalizes that mistake. It does not enable the project team to focus on the constraint to the project but causes them to make the error of chasing an ever changing critical path. As Dr. Deming illustrated with the funnel experiment, that always makes the project system performance worse.

The critical chain does not change during project performance, partly a matter of definition but mostly a result of the overall critical chain plan construction procedure and the subordination step.

4.2.2 Exploiting the constraint

Having defined the critical chain as the constraint to performing the project faster, you should now look to exploit the constraint. That means reducing both the planned time and the actual project performance time. CCPM exploits the critical chain using an understanding of variation. This is where Dr. Goldratt's unique focus on statistical fluctuations and dependent events leads to a significant departure from most current project systems. Dr. Goldratt's recognition of variation is not unique; but his solution applied to the project management system is innovative.

Dr. Deming noted that managers often make many systems worse by not understanding the fundamental difference between common cause and special cause variation. He also notes, "I should estimate that in my experience most troubles and most possibilities for improvement add up to propositions something like this: 94% belong to the system (responsibility of management), 6% special."

It is not news that projects have common cause variation in the performance time of tasks. Although the time to perform individual project tasks may be independent of each other, project task networks define task dependence. By the definition of the project logic, the successor task cannot start until the predecessor task is complete (for the most frequent finish-to-start task connection).

The TOC improvements for production take advantage of (exploit) the reality of statistical fluctuations and dependent events. Section 3.3.1 described how concentrating contingency at the end of the critical chain accomplishes that.

Common cause variation in task performance is not an exceptional event, such as discrete project risk events. PERT attempted to estimate the impact of common cause variation by using three task duration estimates but for a variety of reasons did not succeed. The PMBOK™ Guide and literature still make mention of PERT in this fashion, although it is little used today. PERT diagrams, as referred to in much of the project literature and in many project software packages are simply a way to show the project network logic independent of the time scale, not an application of the three time estimates. Some projects use methods such as simulation and Monte Carlo analysis to assess the impact of task duration and cost uncertainty. While those methods propose a way to estimate uncertainty, they do not pose an effective systematic method to manage it.

CCPM accounts for common cause variation as an essential element of the project management system. The process removes identifiable special causes of variation, including resource unavailability and common resource behavior patterns, including the student syndrome and multitasking. CCPM project managers use resource buffers to identify and ensure availability of resources on the critical chain.

4.2.2.1 Exploiting project task estimates

CCPM seeks to use best estimate, or 50% probable, individual task time estimates. The CCPM project manager recognizes that actual individual task performance times include common cause variation and does not criticize task performers for individual task duration performance.

Chapter 3 noted that most project managers attempt to account for individual task common cause variation by adding contingency time into each estimate. (Recall from Chapter 3 that the operational definition of *contingency* in this book is the difference between the 90-95% probable estimate and the 50% probable estimate.) They usually do not specify the existence or amount of that contingency time. People estimating task times for a project usually do so believing that the project manager wants low-risk task times, perhaps a probability of 80% to 95% completion on or less than the task duration estimate. Figure 3.2 illustrated that that estimate is two or more times the 50% probable estimate. In most project environments, people feel good if they complete a task by the due date and feel bad if they overrun the due date. That reinforces their attempts to estimate high-probability completion times.

Deming's mentor, Walter A. Shewhart, made the following observation [2]:

> It should be noted that the statistician does not attempt to make any verifiable prediction about one single estimate; instead, he states his prediction in terms of what is going to happen in a whole sequence of estimates made under conditions specified in the operational meaning of the estimate that he chose.

That view clarifies why attempts to deal with uncertainty for individual task estimates are fruitless.

Some experienced project managers say that people tend to give optimistic estimates (i.e., too short). They base that contention on remembering the instances in which projects had difficulty meeting the delivery date. Generalizing the observation does not hold up under examination for several reasons.

First, extensive psychological research demonstrates that people tend to seek pleasure and avoid pain. In most project environments, people get pleasure and avoid pain by completing tasks on the due date. Hardly anyone wants to be known as the person who can be counted on to deliver late. It is not reasonable to expect people to solicit pain by systematically giving "optimistic" estimates.

Second, people remember selectively. They easily remember worst-case outcomes (pain) but not necessarily all the times things went to their advantage. Most people feel that they always pick the slowest line in a bank or supermarket, but could that really be true? People also tend to forget predecessors leading to the outcome (as in the student syndrome), a mental feat that has two interesting effects:

▶ Project managers selectively remember the instances in which task duration estimates were exceeded and therefore want to add contingency of their own.

▶ Task performers tend to add time to their next estimate.

Third, if underestimating task durations were the predominant fact, nearly all projects would be late. Assuming that most of the potential positive variation in task times is returned to the project (evidence suggests otherwise), the merging of task paths ensures a very low probability of success if individual estimates are less than 50% probable. (Real project behavior is, of course, confounded by control actions taken during project performance. Those control actions may help or hinder overall completion time performance.)

While many projects do fail to meet schedule, our observations indicate that a substantial portion achieve the scheduled project end date. Almost all projects to create bid proposals complete on time. Nearly all major meetings come off as planned with few problems. The Olympics have not yet been delayed because of late project completion. (The stadium in Atlanta caused anxious moments but was ready nonetheless.)

Milestone performance in one large project demonstrated that the task performance data conformed closely with Dr. Goldratt's prediction that about 80% of the task milestones are achieved exactly on schedule, with only one or two sooner and the rest later, including a few significantly later. The project consisted of about 30 large subprojects, some of which contained smaller subprojects.

My experience shows project plans from a variety of organizations (numbering in the hundreds) either fail to specify what probability and confidence of estimate is expected for task duration estimates or fail to provide a quantitative basis for the estimate. The PMBOK^{TM} Guide admonishes project managers to provide those estimates but provides little guidance on what to do with them. Construction projects are somewhat of an exception, having access to extensive quantitative data. For example, the *National Construction Estimator* [3] uses an extensive database and lists many potential contributors to common cause uncertainty in the estimates. The guide states that many of these uncertainty items have ranges of several tens of percentage points of the cost estimate. Therefore, in many cases, they have the same potential impact on schedule.

4.2.2.2   Exploiting statistical laws governing common cause variation

Chapter 3 described how CCPM can and does exploit the statistical law of aggregation by protecting the project from common cause uncertainty of the individual tasks in a task path with buffers at the end of the path. Buffers appear as tasks in the project plan but have no work assigned to them.

4.2.2.3   Exploiting resource availability

One of the leading alleged causes of late projects is that resources are not available or not available in sufficient quantity when they are needed. CCPM requires a mechanism to prevent the critical chain tasks from starting late or taking longer due to the resource. The selected method is to use a resource buffer to provide information to the critical chain resources about when they will be needed.

The resource buffer is different from the project buffer and feeding buffers in that it does not occupy time in the project network. It is an information tool to alert the project manager and performing resources of the impending necessity to work on a critical chain task. Note that inherent in the critical chain idea is that you cannot deterministically schedule resources. Because each task performance will vary, any forward deterministic schedule is an uncertain estimate.

You establish the lead time necessary for each resource on the critical chain of the project and use the project measurement and control process to alert the resource as the time of actual task performance approaches. You can use multiple notifications (e.g., at one month, one week, and two days), different times for different resources, or a standard time.

You may choose to use alternative methods for subcontracted resources, such as contract rewards or penalties for delivering to a specified lead time or duration.

4.2.3 Subordinating merging paths

Most projects have multiple task paths. All task paths must merge into the critical path by the end of the project, if for no other reason than as a milestone that identifies project completion. Usually, the path merges tend to concentrate near the end of the project. One reason is that assembly or test operations tend to occur near the end of a project, requiring many elements to come together. The following statement demonstrates how that becomes a primary cause of the well-known project truth: "Many projects complete 90% in the first year, and complete the last 10% in the second year." Figure 4.5 illustrates the filtering effect of merging paths. The successor task cannot start until the latest of the predecessor tasks is complete.

The merging of task paths creates a filter that eliminates positive fluctuations and passes on the longest delay. The reason is that merging task paths means that all the feeding paths are required to start the successor task. Therefore, the successor task cannot start until the latest of the merging tasks completes. Consider a task on the project critical path that requires three separate inputs in order to start. That is frequent in assembly operations and in many project results, such as a major show or meeting event where everything has to be ready on opening day. Usually, there are many more than three inputs. However, even with three, if each has a 50% chance of being done in the estimated time, the probability that

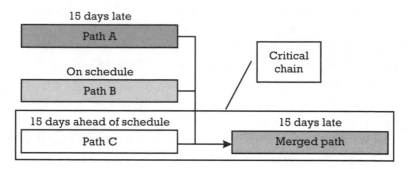

Figure 4.5   Merging paths cause critical chain delay if any of the feeding chains is delayed.

at least one is late is over 88%! Even if each individual task had a 90% probability of completion, the probability of at least one being late is still 30%, or nearly one out of three.

CCPM protects the critical chain from potential delays by subordinating critical chain feeding paths: placing an aggregated feeding buffer on each path that feeds the critical chain. Figure 4.6 illustrates the placement of the feeding buffers. It includes paths that merge with the critical chain at the end of the project. The feeding buffer provides a measurement and control mechanism to protect the critical chain. The figure also illustrates how the buffers absorb the late paths.

That innovation immunizes the critical chain from potential delays in the feeding paths. It also provides a means to measure the feeding paths, while keeping focus on the critical chain.

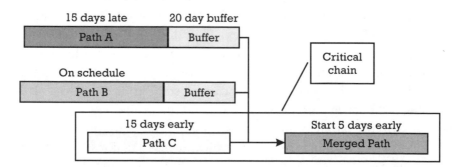

Figure 4.6   Feeding buffers absorb fluctuations in critical chain feeding paths.

4.2.4 Task performance

4.2.4.1 Elevating date-driven performance

The primary local optimum of significance in project management is the estimate of each individual task time. If management judges the performer of each task based on completing their task on the estimated milestone date (local optima), what does that do to the overall project completion time (system optima)? Section 3.2.2 described the phenomena that work to ensure that few tasks get reported as done early. Most get reported as done on schedule.

Critical chain project plans provide dates only for the start of task chains and the end of the project buffer. For the rest of the project, the plan provides approximate start times and estimated task duration. Critical chain project managers do not criticize performers who overrun estimated task durations, as long as the resource (a) started the task as soon as they had the input, (b) worked 100% on the task (no multitasking), and (c) passed on the task output as soon as it was completed. They expect 50% of the tasks to overrun.

4.2.4.2 Elevating task performance by eliminating multitasking

Multitasking is the performance of multiple project tasks at the same time. Some people refer to it as the fractional head count; it is much like rubbing your stomach and patting your head at the same time. People actually multitask by dividing out time between the multiple tasks. People might do this during the course of the day by working on one project in the morning and one in the afternoon.

Most people think of multitasking as a good way to improve efficiency. It ensures everyone is busy all the time. Often, I have to wait for inputs or for someone to call back before I can get on with a task. Multitasking makes good use of that time.

Dr. Goldratt demonstrated in *The Goal* how focus on local efficiency could damage the overall performance of a system. He used the example of robots operated all the time to show high efficiency. In the case of production, that leads to producing excess inventory and may plug the constraint with work not necessary for current orders, increasing operating expense and delivery times with no positive benefit to the company as a whole.

If multitasking is a normal way of business in a company, three weeks becomes the normal task duration for a task. Performance data support that inflated task duration. If this is a critical chain task, the practice directly extends the duration of the project. Most companies admit to encouraging extensive multitasking.

CCPM seeks to eliminate that type of multitasking by eliciting 100% focus on the project task at hand by all resources supporting the project. Thus, eliminating fractional head counts is a primary consideration in planning a critical chain project. Eliminating resource contention within the plan eliminates the pressure to multitask on a single project.

I am often asked, "Isn't it a manager's job to multitask?" or "What if I am held up on one project task?" My answer is to clarify that there can be good multitasking. Bad multitasking is multitasking that extends the duration of a project task. As long as you position yourself and your project work to avoid bad multitasking, you are contributing your best to the project team.

4.2.5   Early start versus late finish

Extensive studies have evaluated the desirability of using early start schedules or late finish schedules. Project managers believe that early start schedules reduce project risk by getting things done early and that late finish schedules accomplish the following:

‣ Reduce the impact of changes on work already performed;

‣ Delay the project cash outlay;

‣ Give the project a chance to focus by starting with fewer simultaneous task chains, allowing the project team and processes to come up to speed.

Much project management guidance recommends that project managers use an early start schedule. Many schedule computer programs use the early start schedule as the default. Early start means permitting all the noncritical path tasks to start earlier than is necessary to meet the schedule date. People working on those tasks know there is slack in their task. How does that influence the urgency they feel in working on the task? Does it encourage or discourage the student syndrome?

CCPM uses late start for all project tasks. Note that the feeding buffers provide an explicitly sized buffer to protect the overall project from late completions in the feeding paths. That maximizes the advantages to the project, while ensuring project schedule protection.

I am often asked, "Yes, but what does it hurt to start early if I have the resource?" I answer by agreeing that once you understand this theory, if it does not hurt anything, by all means do it. TOC requires that people use their knowledge and intuition.

4.3 Exploiting the plan using buffer management

Measures drive actions that move you toward the goal. In *The Haystack Syndrome*, Dr. Goldratt notes [4]:

> The first thing that must be clearly defined is the overall purpose of the organization—or, as I prefer to call it, the organization's goal. The second thing is measurements. Not just any measurements, but measurements that will enable us to judge the impact of a local decision on the global goal.

Figure 4.7 illustrates the cybernetic view of measures used by Dr. Joseph Juran. The sensor makes the measure in block 2. An umpire

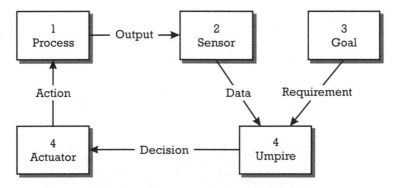

Figure 4.7 Dr. Joseph Juran depicts measurement as part of a control system.

(block 4) compares the output of the process as reported by the sensor to the goal for the process. The umpire makes a decision to cause an action, modifying the process to change output and minimize the gap. That is how all control systems work and is the intent of project measurement systems, in which the goal includes the technical requirements, cost, and schedule for the project.

In *The Haystack Syndrome*, Dr. Goldratt defines data as "every string of characters that describes something, anything, about our reality" [4]. He defines information as "the answer to the question asked." Dr. Goldratt suggests that the information system should incorporate the decision.

The improved measurement system for CCPM follows the practice established by Dr. Goldratt for production operations. It uses buffers (i.e., time) to measure task chain performance. Recall that the end of the project buffer is a fixed date: the project delivery date. For buffer management purposes, you also fix the ends of the feeding buffers. You determine project buffer penetration by asking people working on tasks, "When will you be done?" That allows you to project forward using the downstream task duration estimates to predict how much of the buffer would be used up if they complete at that time.

You size the buffers based on the length of the task chain they project. Buffer sizing uses the uncertainty in the duration of the critical chain tasks to size the project buffer. Likewise, uncertainty in the duration of the feeding chain tasks determines the size of each critical chain feeding buffer (CCFB). CCPM sets explicit action levels for decisions. The decision levels are in terms of the buffer size, measured in days:

1. Within the first third of the buffer: Take no action.

2. Penetrate the middle third of the buffer: Assess the problem and plan for action.

3. Penetrate the final third: Initiate action.

Those measures apply to both the project buffer and the CCFBs. Figure 4.8 shows an example of using the buffers. The three Xs show three potential amounts of buffer penetration corresponding to the above criteria.

Project teams monitor the project buffer and each CCFB at the appropriate time intervals for the project, usually weekly but at least monthly.

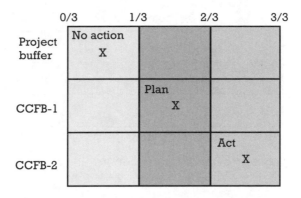

Figure 4.8 Buffer penetration provides action decisions.

You can update the buffers in a relatively short weekly meeting or use E-mail status reports. For this tool to be fully useful, the buffer monitoring time must be at least as frequent as one-third the total buffer time. If the buffers are negative (i.e., the latest task on the chain is early relative to schedule date) or less than one-third the total buffer late (e.g., less than 10 days if the total buffer is 30 days), you do not need to take action. If extended durations penetrate the buffer between one-third and two-thirds, the project team should plan actions for that chain to accelerate the current or future tasks and recover the buffer. If the task performance penetrates the buffer by more than two-thirds the buffer size, the project team should take the planned action. Through this mechanism, buffer management provides a unique anticipatory project management tool with clear decision criteria.

Project managers update the buffers as often as they need to simply by asking each of the performing tasks how many days they estimate to the completion of their task. They do that without pressure or comment on the estimate. They expect the estimates to vary from day to day and some of the tasks to exceed the original duration estimates. As long as the resources are working on the tasks with the CCPM task performance paradigm, managers evaluate them positively, regardless of the actual duration.

An enhancement in the use of the buffer for long critical chains is to plot trends for buffer utilization, as shown in Figure 4.9. The buffer measure then becomes functionally similar to a control chart and can use similar rules. That is, any penetration of the red zone requires action. Four points trending successively in one direction require action.

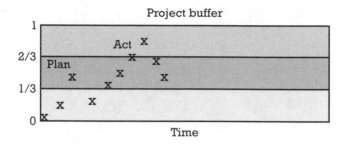

Figure 4.9 Plotting trends of buffer penetration may aid analysis.

People have suggested that project managers also monitor buffer penetration as a percentage of critical chain use. The idea is that the manager should not use up the project buffer too early in the project. Trending buffer penetration has several advantages over that approach. Perhaps the most important advantage is that it is simpler to interpret. A second advantage, especially important if the project task processes are not in statistical control, is that trending preserves the time history of the data. That is important information to help improve control of the processes.

Updating the buffers requires that you maintain project status versus your plan in terms of the tasks completed. It is also a useful direct measure of project performance.

4.4 Features (more or less) from PMBOK™

The unique features of CCPM do not make up a sufficient system to satisfy the project system requirements identified at the beginning of this chapter. The PMBOK™ Guide seems to provide all the necessary additional features to meet the complete system requirements. Following Juran, I have prepared a feature and requirement correlation matrix to examine how the CCPM features and selected features and processes from the PMBOK™ Guide combine to provide the complete set of identified project system requirements. (The table is too large for publication in book format; see [5].) It helped to identify the following set of PMBOK™ Guide features as the primary ones necessary to deliver to the requirements given in Table 4.1. The correlation table also leads to clarification of the requirements that pertain to each feature and therefore supports developing the feature.

The following features, which are (mostly) contained in and explained in the PMBOK™ Guide are necessary to satisfy these requirements:

4.4.1 Project charter

The project charter authorizes the initial project team to prepare the project work plan. It identifies the overall project deliverable, project stakeholders, overall project responsibilities, and other parameters necessary to create an effective project work plan.

4.4.2 Project work plan

The project work plan identifies the scope, budget, schedule, responsibilities, and resource requirements for the project. It may also specify other project requirements and plans to achieve them, such as quality, safety, and regulatory plans. It must contain or reference the operational procedures for the project. Key elements of the project work plan include the WBS, responsibility assignment, milestone sequencing, work packages, and the project network.

4.4.2.1 · Work breakdown structure

The WBS is the framework to define project scope. It defines project scope hierarchically, from the complete project level to the work package level. Work packages complete the hierarchy by specifying the project tasks necessary to deliver the scope.

4.4.2.2 Responsibility assignment

Responsibility assignment designates the individuals responsible to accomplish deliverables on the WBS. Responsibility assignment must occur at the work package level and may be assigned at higher levels. Responsibility assignment normally confers the authority to perform the work and accountability for delivering the scope to the budget and schedule for the project deliverable. The person responsible for a work package does not have to be a task performer, that is, a resource on a task.

4.4.2.3 Milestone sequencing

Milestone sequencing is a tool to go from the hierarchically formatted WBS to a logical project plan. It provides the major sequence of project tasks for use by work package managers to link the inputs and outputs of their work packages. (This element is not described in the PMBOK™ Guide, but it is covered in Chapter 5.)

4.4.2.4   Work packages

Work packages define the plan to produce project deliverables at the lowest level. Work packages contain the scope definition for the deliverable of the work package and the plan to produce the deliverable. This plan includes defining the project tasks, the logic for the tasks, and the linkage of the work package tasks to other elements of the work plan; usually to milestones on the milestone sequence chart. Work packages may link to tasks in other work packages as well, but this linkage usually cannot occur on the first draft because all work packages are planned simultaneously. Work packages also identify the estimated task duration and resource requirements and the assumptions necessary to support those estimates.

4.4.2.5   Project network

The project network logically connects all the tasks necessary to complete the project. The project tasks must identify the resources necessary to perform the task within the estimated task duration. The network includes all the tasks from all the work packages, and identifies the critical chain, the project buffer, and the CCFBs. It provides start dates for each chain of tasks and the completion date for the entire project. It is the basis for subsequent performance measurement and control.

4.4.3   Project measurement and control process

CCPM defines an improved schedule measurement and control process. Most projects also require a technical quality control process, and many projects also require a cost control process.

The correlation matrix [5] also identified a need for processes to ensure project result quality and provide mechanisms for continuous improvement. The scope of this text does not address the process to ensure project quality results. Ireland [6] provides an overview of a satisfactory process to meet those requirements.

4.4.4 Project change control

The project measurement and control process will, from time to time, trigger the need for action to complete the project successfully. In addition, unfulfilled assumptions made at the start of the project, as-found conditions that differ from initial assumptions, or changes in the client's demands may require changes in the remainder of the project. Project change control defines a process to incorporate and communicate those changes to all members of the project team.

4.4.5 Project risk management

Project risk management handles potential causes of special cause variation. Because the PMBOK™ Guide does not differentiate between common cause variation and special cause variation, it addresses both under the realm of project risk management. CCPM addresses common cause variation directly in the project system and thus confines project risk management to special cause variation. (See Chapter 11.)

4.5 Summary

This chapter developed the project system requirements and described the critical chain system features and key supporting PMBOK™ system features designed to satisfy those requirements. Key system features are as follows:

- ▶ The critical chain identifies the project constraint.

- ▶ Exploiting the critical chain utilizes uncertainty management in the form of reduced task plan durations and a project buffer.

- ▶ Exploiting the critical chain requires resource buffers.

- ▶ Feeding chains and resource efficiency are subordinated to the critical chain with CCFBs.

- ▶ Critical chain projects rely primarily on buffer management for project control.

- ▶ Additional features from the PMBOK™ Guide are necessary to complete an effective project management system.

These high-level system features are necessary and sufficient to satisfy the requirements.

References

[1]   Juran, J. J., *Juran on Planning for Quality*, New York: The Free Press, 1988.

[2]   Shewhart, W. A., *Statistical Method from the Viewpoint of Quality Control*, New York: Dover, 1986 (originally published in 1939).

[3]   Kiley, M. D., *1997 National Construction Estimator*, Craftsman Book Company, 1996.

[4]   Goldratt, E. M., *The Haystack Syndrome*, Croton-on Hudson, NY: North River Press, 1990.

[5]   www.Advanced-projects.com.

[6]   Ireland, L. R., *Quality Management for Projects and Programs*, Upper Darby, PA: PMI, 1991.

Starting a new project

5.1 Project initiation process

The project initiation process ensures that all the conditions necessary for project success are met. It starts with an agreement by all the project stakeholders on the expected project results and ends with a clear understanding of who is responsible and accountable to do what and by when to achieve those results.

Figure 5.1 illustrates the overall process to successfully initiate a project. It starts with the project charter, an often overlooked but necessary part of any project. It ends with a project work plan that is sufficient to start work on the project.

The PMBOK™ Guide [1] considers the project initiation process in a more restricted sense than does this chapter. It identifies the outputs of project initiation as project charter, project manager assignment, project constraints, and project assumptions. While this text includes all those items, we extend project initiation all the way through initiating project work to an effective work plan.

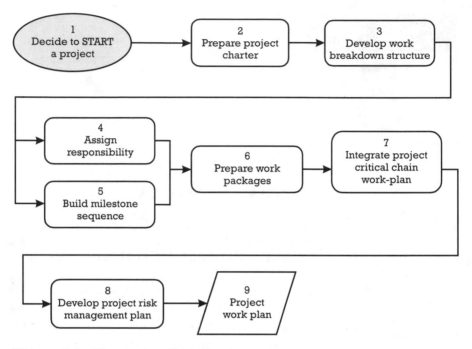

Figure 5.1   The project initiation process.

5.2   The project charter

A project charter is a brief written statement to enable the assembly of an effective team to plan the project. The definition goes well beyond the charter described in the PMBOK™ Guide, including summarizing the who, what, when, where, why, and how. It normally should include all the elements described by CH2MHILL [2] as essential:

- ▶ Vision;

- ▶ Purpose;

- ▶ Membership;

- ▶ Mission;

- ▶ Organizational linkage;

- ▶ Boundaries;

- ▶ Team and individual responsibilities;

▶ Measures of success;

▶ Operating guidelines.

The primary distinction to make between the project charter and the project work plan is that the charter authorizes development of the work plan.

5.3 Stakeholder endorsement

There is a difference between being involved and being committed. Consider a bacon and egg breakfast: The chicken is involved; the pig is committed. The idea of endorsement is to get everyone who may have an impact on your project committed to your plan at the beginning. All too often, project participants with a direct effect on project success, such as the customer or senior management, get the idea that they do not have a key role in creating project success and instead set themselves up in a role to judge rather than create. That is a high-probability precursor to project failure. The project team must ensure that all parties that have a potential impact on project success endorse the project to the degree necessary to ensure project success. There are many ways to accomplish that. One requirement to obtain endorsement is to be sure that your team has listened to and addressed the needs of each of the stakeholders. Usually, you should obtain formal endorsement of both the project charter and the project work plan. In some instances, the project contract may help to fulfill one of those roles. If you do not have a signed contract for your project, it is a good idea to get the client to be an "author" for the project charter and the project work plan, even if you write it for them.

5.4 The work breakdown structure

The WBS provides a common framework for planning and controlling the work to be performed. It provides a common, ordered framework for summarizing information and for quantitative and narrative reporting to customers and management. The WBS uses a hierarchical breakdown of project deliverables. This breakdown is intended to provide more manageable pieces of work and a framework for overall operation and control of the project.

Critical chain project management texts suggest using a TOC tool, the PRT (see Figure 11.5 for an example), to create the WBS [3]. The idea is to start with the end item or an intermediate objective and ask the team, "What obstacles prevent us from achieving this objective?" Once you have the list of obstacles, ask the team to state conditions that would overcome the obstacles. Then link those conditions in a logical sequence. This method ensures a coherent strategy and synchronized tactics to overcome the obstacles identified by the team. For very large projects, you could create layers of PRTs, corresponding to layers in the WBS. Unfortunately, the simplified method for creating the PRT described by Goldratt in *It's Not Luck* [3] is not appropriate to generate a project WBS. The reason is that the PRT only ensures that certain necessary conditions are met, those necessary to overcome the obstacles that the team identifies. The method does not ensure inclusion of all deliverables sufficient to deliver the project result. Dettmer describes a modified method to ensure both necessity and sufficiency [4].

Kerzner provides the following criteria for a WBS: "The project manager must structure the work into small elements that are manageable, in that specific authority and responsibility can be assigned; independent or with minimum interfacing with and dependence on other ongoing elements; integratable, so the total package can be seen; and measurable, in terms of progress" [5].

A properly prepared WBS should facilitate the following:

▶ Ensuring better understanding of the work;

▶ Planning of all work;

▶ Identifying end products and deliverables;

▶ Defining work in successively greater detail;

▶ Relating end items to objectives;

▶ Assigning responsibility for all work;

▶ Estimating costs and schedules;

▶ Planning and allocating company resources;

▶ Integration of scope, schedule, and cost;

▶ Monitoring cost, schedule, and technical performance;

▶ Summarizing information for management and reporting providing traceability to lower levels of detail;

▶ Controlling changes.

The WBS usually has levels assigned, for example:

Level 1: Total program

Level 2: Summary cost accounts

⋮

Level $n - 1$: Work package

Level n: Activity

In some cases, these terms have different meanings. In particular, in many cases the work package is the lowest level of work assignment, restricted to one resource provider per work package.

The WBS also has a numbering system that provides a unique number to every piece of work defined. The numbers usually follow the hierarchy of the levels, with the lowest level corresponding to a charge number for collection of cost.

Project managers use different approaches to subdivide a total project into a WBS. The most preferred is a product-oriented WBS (as shown later, in Table 5.1), in which each work package produces a definable, measurable output. The collection upward then may follow functional lines or, for major pieces of hardware (including facilities), subsystems and systems.

The most important aspect of the WBS is that it be comprehensive. Because it is the basis for all planning and cost estimating, nothing should be left out. Also, if the project funding decision is going to be based on cost, it is imperative that the WBS not be redundant. That is, each activity should appear in only one work package.

Many companies use templates to create WBSs for similar projects, which can be a useful resource to get started. However, templates share a major shortcoming with other checklists in that they tend to provide a degree of comfort and sometimes stifle thinking beyond the items in the checklist. The project manager has to be vigilant not to allow templates to constrain the thinking to ensure that all required work is covered in the WBS.

Sometimes clients (especially government clients) will dictate a WBS structure, usually because they have a need to compare across projects by different contractors or for different types of purchases. That is a legitimate client need and must be honored. The project manager still must ensure that all project work is covered, that there are no redundancies, and that responsibility assignments are unique and appropriate.

Do not confuse the WBS with the project or company organization structure. Although work may align with the organization, it does not have to. The only requirement is that at the lowest level, one individual has clear responsibility and authority for the work performed. More important, the WBS must define the deliverables for the project, not the functions necessary to deliver the scope.

There are many opinions on how to organize a company for project management. Because most project managers do not have the luxury to redesign the company for their project, we will not address the overall company organization. Project managers usually do have the flexibility and authority to design their WBS, select their project management team, and assign responsibility and authority.

Following Dr. Deming's idea to use the overall process flow for a company as an organizing principle, an alternative we recommend is that the project team be organized around the WBS. Another alternative is to place someone responsible for the critical chain and for each of the CCFBs. Since it is unlikely that the WBS was organized that way, the project management team may cross-cut the responsibilities of the work package managers. That places the project management team responsibility on the connections between work packages and activities, the most vulnerable part of a project.

5.5 Responsibility assignment

Responsibility assignment ensures that every element on the WBS is owned by someone. It used to be the fashion to create a responsibility assignment matrix, which places the WBS on one side and the organization structure orthogonal to it. The matrix is sparse if you designate only the person responsible for the specific WBS element, and for an organization of reasonable size it is too large for people to handle. It also is hard to use. Finally, a responsibility assignment matrix is difficult to change if a company changes its organization.

A superior representation is the linear responsibility matrix, which lists the WBS elements in the first column, the responsible person (not organization element) in the second column, and anything else you want in subsequent columns. The linear responsibility matrix is easy to develop and maintain. You can look at it on a computer screen, print it on regular paper, and bind it in the plan so everyone can use it. It also can convey much more information. Table 5.1 is a simple example of a linear responsibility matrix.

5.6 Milestone sequencing

The WBS defines the scope of the project deliverables and the key processes necessary to provide the deliverables (e.g., design), but it provides no information on the sequence of project tasks. The project plan must logically sequence all the project tasks. For a small project (i.e., 50 tasks or fewer), you can go directly from the WBS to a task list and link the tasks using project scheduling software. For a project with a larger number of tasks, that approach does not work. The number of task linkages rapidly becomes too large to link even a WBS-ordered task list. You need an intermediate step to facilitate generating the project task logic.

An effective way to aid developing the logic is first to identify the major project phases, in terms of key milestones for the project. Figure 5.2

Table 5.1
Example of a Linear Responsibility Matrix

WBS Number	Deliverable	Responsible Person	Notes
1	Design package	Karl Sagan	—
1.1	System engineering	Karl Sagan	Lead for integration design reviews
1.2	Hardware design	Charles Metcalf	—
1.3	Software design	Simon Ligree	—
2	First prototype delivered	Mary Riley	—
3	System tests	John Jones	—

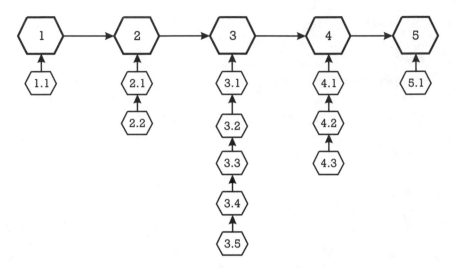

Figure 5.2   The key milestones define a "backbone" for project task sequences.

illustrates an example of the structure of a key milestone chart. Each milestone must have a specific deliverable assigned to it. The milestone sequence chart does not include dates. Dates result from the integrated schedule; they are not inputs to it unless it is a project with a definitive end date, such as a proposal submission or a meeting, such as the Olympics.

Then ask what is necessary to complete each of the major milestones and build a list of supporting milestones under each key milestone. The resulting milestone sequence chart, worked jointly by all the key project team members, provides a basis for developing and sequencing the tasks defined in the work packages. It provides many of the linkage points to tie the work packages together.

You can use the milestone sequence as a supplemental tool for project measurement. Many organizations establish project decision gates as key points for project reviews, such as completion of the system engineering or completion of the first prototype test for development projects or completion of the conceptual design for construction projects. Expect to find those major milestones on the critical chain for the project.

Management and clients often like to use milestones as indicators of project success on performance to schedule. If you put the milestones on the critical chain of the project plan, there is a very low probability that they will be completed on time; therefore, performance is subject to mis-interpretation by people who do not yet understand the critical chain process. In that case, we recommend adding a project buffer to each major measurement milestone and reporting using the end of the milestone buffer as the milestone commitment date. Figure 5.3 illustrates that idea. You should still control the project to the overall project buffer.

5.7 Work packages

Work packages provide the basis for the project network, schedule, and cost estimate. They are contracts between the project manager and the work performers. They are the source documents for inputs to the integrated cost schedule plan for the project. Work packages contain the scope to be delivered, specifications or reference to specifications, codes, and standards for the deliverables, the activity logic, activity resource estimates, and the basis for the activity resource estimates.

The design of work package documentation can greatly influence the ease and quality of the project planning. It is the point at which most engineers begin to complain about too much paper. The design of the work package process must be simple and user friendly. Figure 5.4 illustrates the project logic input, an essential part of the work package. This representation includes task title, duration, and resource requirements on the logic sketch. Combined with the assumptions and deliverables (scope statement), the logic sketch (subnet or "fragnet") delivers the information necessary for a project plan.

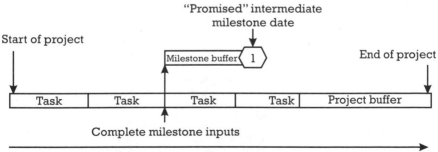

Figure 5.3 If your organization uses milestones to judge project progress, you must put a buffer in front of them.

Work package WBS number:_____   **Work package deliverables:**
Work package title:_____
Work package manager:_____
Date:_____   **Work package assumptions:**

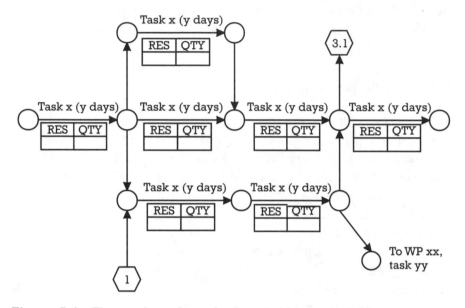

Figure 5.4   The work package logic provides essential input to create the project plan.

Elements on the WBS are assigned to people to plan and manage. Those individuals usually have a title, such as work package manager or cost account manager. They are usually technical experts in the subject matter of that portion of the WBS. They must define the detailed work scope, establish the task sequence, and estimate the task resource requirements. They are responsible for identifying the links between their work packages and other work packages in the program. They also supply the justification for the resource estimates.

5.7.1   Assumptions

Every project plan is based on assumptions. No matter how much detail you put into the project specifications, there is always a lower level of assumptions underlying that detail. Project plans should identify the assumptions necessary to provide reasonable estimates of the project task parameters: resources required and task duration. For example, an assumption for a construction schedule may relate to the weather: "No more than six days of outside work lost due to inclement weather." An assumption might address actions outside the direct control of the project, for example, "Permit review time by regulatory agency not to exceed 30 days." Identify those assumptions during development of the work packages.

Try to counter two frequent tendencies in writing assumptions. One tendency is to assume everything rather than do the necessary planning work. That can lead to long lists of assumptions, which can be summarized as, "We're not responsible for anything." Limit your assumptions to those necessary to create an effective plan. Another tendency is to write assumptions in the negative, that is, to specify what the project will not do or what is not in the scope of the project, rather than specify project deliverables. A better tactic is a positive general statement that says, "Project deliverables include only the specified items."

5.7.2 Project logic

The most important part of the work package documentation is the network input. Note that the network input does not carry dates. It provides activity duration and logic and ties the resource requirements to the activities. The reason that dates are not put on the network is that the dates cannot be developed until the network is put together and the critical chain developed. I have seen several companies put the project input format into the hands of budget personnel who create forms that are budget request spreadsheets. Planners have to separately develop the schedule and figure out how to make them match. Use the work package as designed and let the computer determine the schedule and spread the budgets for you. Task dates and budget spreadsheets are outputs from the integrated cost schedule system, not inputs. (Of course, in critical chain we are not interested in most of the task dates anyway; all we are interested in is the start date of chains of tasks and the completion date of the project.)

The project logic defines the necessary sequence of tasks to achieve the project result. Work packages are simply small projects; each requires its own logic. You must link tasks so the output of one task is the input of the next task. The most common task logic links the finish of one task to the start of the next. The most common relationships or links are:

▶ *Finish-to-start* (FS). Often called the predecessor-successor relationship, it clearly illustrates how the output of one task is required as the input to start the next task in the sequence.

▶ *Start-to-start* (SS) (with a lag). Use this relationship when two tasks can be carried out simultaneously, once the first task has created some amount of output for the second task to work on. For example, you may have to create many copies of something that requires three steps. Rather than schedule all three steps for every copy, you put in one task for each of the steps titled something like "Step 1 for 100 copies of *x*, step 2 for 100 copies of *x*," and so on, with each step lagged by the amount of time necessary to complete the first item.

▶ *Finish-to-finish* (FF). Use this relationship when things must end up at the same time but may have different start times.

Most computer scheduling software offers a host of other possible constraints, including fixed-date constraints. Use such other constraints as little as possible. Dates should be an output of your network, not an input to it.

Check your project logic, considering the following points:

▶ Does each task have a clearly defined output?

▶ Are the predecessors to each task necessary to start the task?

▶ Are the predecessors to each task sufficient to perform the task?

▶ Do the tasks (collectively) provide for all the project deliverables as outputs (compare to the WBS)?

▶ Do the tasks specify the necessary resources?

▶ Do the tasks have unnecessary date constraints?

▶ Are all the milestones on the milestone sequence chart included?

▶ Are the resources that determine task duration working at 100% utilization?

▶ Do all the project network paths tie in to the end of the project? (If not, tie them in at least to a milestone for "Project Complete.")

Chapter 6 describes how to create the critical chain plan from this resource-loaded project network.

5.7.3 How many tasks?

Many project plans contain thousands of individual tasks. Only rarely should project plans exceed a few hundred activities. Guidance on how many tasks to include usually suggests breaking down tasks to very small time durations, to facilitate project reporting by task completion rather than using estimates of percentage completed or time remaining. Consideration of the purpose of the project plan and TOC leads to recommending that project plans be as detailed as you need to successfully run your project, but no more. Project plans are not a substitute for detailed design information. You can always have a hierarchy of plans, in which no individual plan contains more than a few hundred activities.

The primary reason to limit the number of tasks in a project plan is that overall uncertainty does not justify too much detail. Too much detail increases the work to create and maintain the plan and the probability of errors in the plan. Few people, even with study, can understand a plan with more than a hundred activities. Because the number of potential links in a plan increases exponentially with the number of tasks, it is highly unlikely that a plan with even a hundred activities will be error free.

Consider the fact that the average size of a task (in terms of dollars, total path time, or total resource time) is the inverse of the number of tasks in the plan. Therefore, the average size of a task in a plan with one hundred activities is 1% of the total project. Since most of the tasks can be estimated with accuracy no better than tens of percentage points, it makes little sense to divide the project into smaller and smaller pieces.

People often suggest that an insufficiently detailed plan is a cause of project failure. When projects do fail, it is usually in the range of many tens, several hundreds, or even thousands of percentage points, not fractions of a percentage point. It is not logical to conclude that plans with one hundred or more activities are not detailed enough to prevent project failure. You are not likely to find missing pieces that add to hundreds of percentage points by subdividing chunks that average only 1% of the total. The problem is elsewhere, as is the solution.

More tasks do increase the amount of effort required to develop the project plan. A given planning effort spread over more tasks means less effort per task. That is more likely to lead to a less accurate plan, not a more accurate plan. If you have the ability to put in more planning effort, you should apply it to looking at the spaces between the tasks and considering the resources and processes within the tasks, rather than adding tasks to the project plan.

For statistical reasons, there is value in ensuring that the critical chain of the plan contains at least ten activities. That increases the chances that statistical fluctuations will tend to offset each other. Also, no single activity duration should exceed about 20% of the critical chain. If one task dominates the critical chain, that task is more subject to variation and more at risk from inaccurate estimates of the time to complete. Consider defining intermediate deliverables to divide a dominant task.

On the other hand, if you have many tasks on a path, and several tasks in sequence use the same resource, consider combining the tasks and defining the final deliverable as the task output.

The above considerations (number and relative size of activities) also apply to feeding chains as well as to the critical chain, but they are less important on feeding chains because the feeding chain is protected by both the feeding buffer and the project buffer.

5.7.4   Activity duration estimate

Chapter 3 demonstrated the importance of the activity duration estimate. When starting with critical chain, solicit task duration as you have always done. Do not ask for the average duration for each activity. People do not have an intuitive sense of average and will tend to give you an estimate they are comfortable with, no matter what you choose to call it. If you ask for the "average," you will have trouble getting a shorter estimate to represent the average.

Make sure that all work estimates include 100% effort on the task. If they do not, reduce the task duration, keeping the work (person-hours or person-days) the same. In other words, if the task had an engineer at one-half time for 10 days, reduce the duration to 5 days, with the engineer working full time.

Only after you have the initial "normal" estimates should you go back and request "average" estimates, using a question like, "How quickly could you perform this task if you had all the inputs you needed at the start and if everything went right?" If you do not get a significantly reduced estimate, you must work with the estimators to understand their reasons. You need a difference between the average estimate and the low-risk estimate to generate the project buffer. Adding a project buffer to low-risk estimates needlessly extends a project schedule.

5.7.5 Uncertainty revisited

Project managers face a conflict over uncertainty of estimates. One pressure comes from management or a client who says, "If your uncertainty is over X%, you must have not done a good job estimating." (Who knows what basis they have for this?) Human beings are by nature inclined to overconfidence in their predictions.

The discussion here treats uncertainty in cost and task duration as if they are interchangeable, which they are when the work is performed by people or resources (e.g., rented machines) charged on a unit or time basis. That is, cost equals the rate ($ per unit) times the number of units. Therefore, uncertainty in the cost (%) is the same as the uncertainty in the use (%). Likewise, because the duration is the number of units multiplied by the time per unit, the uncertainty in the duration (%) is directly proportional to the uncertainty in the cost (%).

Because projects are, by definition, one of a kind and first of a kind, we rarely have statistical information to quantify uncertainty in estimates or task performance. Consider what we do know about uncertainty. Ask someone to give you an estimate on a new house. He or she might start by saying something like, "What are your specifications?" Permanent houses in the United States today range in price from $30,000 to millions of dollars. The most important question about a house is, "Where is it?" Second most important is, "How big is it?" Even with those specifications fixed, prices per square foot can range over a factor of 2, depending on type of construction, interior finishing, and so on. And then, the price can vary by at least 10% for houses with identical specifications, location, and condition, depending on how much the seller has invested, how good a negotiator the buyer is, the general market in the area, the seller's motivation, and other factors.

So we cannot get too close on a house. How much does a car cost? Same routine. Even absolutely identical new cars can vary by at least ±10% for two purchasers, in the same town, at the same time.

A best-selling project management book (which I choose not to reference out of charity) says, "The first type of estimate is an order-of-magnitude analysis, which is made without any detailed engineering data. The overall analysis may have an accuracy of ±35% within the scope of the project." (And I thought "order-of-magnitude" means a factor of 10.) After a few intermediate steps, that text states, "The definitive estimate, also referred to as detailed estimate, has an accuracy of ±5%." Wait a minute. We just agreed that the actual cost of identical existing and very well-known items, automobiles, can vary by twice that amount at the same time and place. How could we possibly expect an estimate of a lesser known entity to have twice the accuracy?

Another source [6] claims in a table that for low-risk projects work package estimates have overall uncertainty of 2%; the subtasks, 5%; the task, 10%; the project, 20%; and the program, 35%. That is, the source claims that as we combine individual estimates of lower uncertainty, we get a higher overall uncertainty—a repeal of the laws of statistics.

It is interesting to find through a review of many project management books that the same cost accuracy estimates keep appearing, but they are never referenced to source material. The only source material referenced relates to construction cost estimating guides, which provide some accuracy estimates quite inconsistent with the project and program cost estimates stated. For example, the *1997 National Construction Estimator* states (boldface is in the original):

> Estimating is an Art, not a science. On many jobs, the range between high and low bid will be 20% or more. There's room for legitimate disagreement on what the correct costs are, even when complete plans and specifications are available, the date and site are established, and labor and material costs are identical for all bidders. [7]

Obviously, other projects, such as R&D or information technology (IT) projects can have much higher uncertainty than construction projects with detailed specifications.

Finally, I could find no books that provide an operational definition of the meaning of the accuracy as ±35%. I may think that is the variance, you may think it is the extreme value, or 99% probability number (assuming a normal distribution, which is probably incorrect for cost estimates). If my understanding is correct, it means that according to your definition, the accuracy is ±115%. (Well, perhaps not minus that amount!) Can we be sure we understand the same meaning?

We all know of several large projects that have overrun their initial schedule and cost estimates by two to three times (see Chapter 1, if you need some reminders) and perhaps even some that spent all the project money and then were canceled, with nothing to show for it. While the multibillion dollar government projects inevitably come to mind first, plenty of large commercial projects have the same performance history. Does that mean there is a systematic bias to underestimate?

Research by major construction firms and experience with critical chain projects demonstrate that projects that complete on schedule usually do so within or near the original cost. (That is not true if the schedule was maintained through extensive overtime.) Projects that overrun do not begin to see that they are in trouble until a significant portion (money or time) of the original plan has been expended, usually one-half to two-thirds of the original estimate. This is the phase of a project in which the expenditure rate is at a maximum. Extending the project duration at the maximum rate creates a disproportionately large impact on project cost.

If the project behavior is as was hypothesized in Chapter 2, that is no credit for positive variances; it takes only one late task on the critical path to make a project late. And, because all the money is spent on the tasks as well because of the "use it or lose it" paradigm, the schedule extension should lead to a cost overrun. (Based on the above, it may lead to a very large overrun.) The large cost overruns are not, however, part of the new critical chain paradigm, because the critical chain method explicitly removes the sources of the overruns.

Expect significant differences between the low-risk estimate to perform a single task and the average estimate for that task. If people suggest that difference is only a few percentage points, they do not understand the reality of variation. Explore the basis for the estimate. You should find that the average task duration estimates are on the order of one-half to one-third the low-risk estimates for the individual task duration.

5.7.6 Cost buffer

If cost is important, the project work plan should include a cost estimate, including a cost buffer. Size the cost buffer for the project considering the project risks and accuracy of the estimates. If you are using project software for your cost estimate, keep in mind that you have estimated each task at its 50-50 probability. Expect to use a significant part of the feeding and project buffer. You have to include an estimate for that use, which is the cost buffer.

You are better off using a single aggregated project cost buffer for the same reasons you are better off using a single aggregated project time buffer. You get the statistical advantages of independent estimates and the psychological advantage of not having it associated with specific tasks.

Keep in mind that you are accounting for two kinds of uncertainty:

1. Bias, which could sum for all the activities subject to the bias;

2. Statistical fluctuations, which will sum as the square root of the sum of the squares for all the independent cost elements.

Bias includes the fact that the variations in cost estimates tend to be skewed distributions. That is, most work package managers have been trained to spend all the money in their work package so they do not get less next time. Your estimate of how successful you are at changing that paradigm should influence your estimate of the cost buffer.

Vigder and Kark performed a recent study on software projects and noted: "A number of the projects we investigated were large-scale systems, involving more than four years duration and more than 100 person-years of effort. Without exception, the costs of all these projects were seriously underestimated" [8]. They note the following as some of the reasons:

▶ The complexity of large-scale systems. Complexity does not increase linearly with lines of code, but rather exponentially.

▶ The larger a system and the further into the future its delivery, the more difficult it is to correctly and completely specify all the requirements. Such cost overruns seem to be amplified for large systems.

▶ The longer the duration between initial requirements and delivery, the more likely that there will be changes in the requirements. That can occur due to changing user expectations, changes to the environment in which the system is to be installed, or new project personnel with different views on what the requirements should be.

▶ Long project duration means that technology advances may outstrip the initial requirements.

Most of these factors do not seem to be confined to software projects. You should consider them when estimating the buffer size to cover bias in the estimates.

Based on that information, the bias portion of the cost buffer should rarely be less than 10% of the total estimated cost. Unless you have a historical track record to substantiate estimate accuracy for your specific project and environment, 10% should be the minimum cost buffer size. It often should be much larger.

The buffer size for fluctuations should consider that you rarely will get the advantage of work package underruns (estimated at their average durations), but you should consider the statistical combination of the positive variances. If you have a dominant work package in terms of total project cost and uncertainty, the uncertainty in that work package should size the statistical part of the cost buffer. If your work packages are similar in size and you have several of them, use the square root of the sum of the squares to size the statistical contribution to the cost buffer.

If the customer is dissatisfied with the size of the cost buffer, consider "rolling wave" planning. This method phases the plan, with a higher level of detail and lower level of uncertainty associated with better-known tasks and less detail with more uncertainty for later phases of the project. The rolling wave method adds detail to the future plan periodically, as it is better defined.

5.7.7 Basis for cost estimates

The cost estimate basis is part of the work package documentation. It is an extremely important element of planning for cost-plus contracts and for many government contracts. It is also the subject of most difficulty for many engineers. They have no trouble estimating resource needs. They just cannot seem to tell you how they came up with that particular number, other than providing a meaningless phrase, such as "engineering judgment" or "past experience." Those phrases are meaningless and insufficient. You might start the question by asking the assumptions they used to come up with the estimate.

Professional estimators have no trouble coming up with the estimate basis. You usually get it without asking. They will refer to guides, previous experience (specific), and quotes from vendors or otherwise substantiate the numbers used. It is usually quite simple to provide the basis for any kind of hardware (e.g., 500 feet of 4-inch pipe at $1.85/ft.; ref. Joe's plumbing telephone quote to Jim A, 3.15.96). There are books of cost estimating factors for routine construction, software coding, and other specific types of skills. The point is to define the estimate basis well enough so that later on, if changes are proposed, you can clearly define what was in the initial scope and what was not.

5.8 The project work plan

The project work plan, sometimes called the project plan or project management plan (PMP), puts the elements developed in the preceding sections into a form accessible to all the project participants. It is the key to communication within the project. It may include or make reference to a number of other elements for larger projects, such as:

- Topical plans, such as quality, safety, procurement, staffing, environmental, or systems engineering;

- Project communication guidelines and project reporting, including document distribution and approvals;

- Work procedures;

- Specifications and standards;

- Change control procedure.

For larger projects subject to changes, you must place the key elements of the project work plan into a system that ensures people work only to the latest approved version of the plan. An effective planning and control policy should include some elements of the following planning and control policy.

5.9 A planning and control policy

All programs and projects must be planned and controlled in accordance with a project management plan prepared in compliance with the following minimum requirements:

▶ Planning must clearly relate cost and schedule to the scope of work.

▶ Planning must be based on the current best estimate of the eventual cost, schedule, and scope of work.

▶ Scope, schedule, and cost estimates must be written. The basis for the estimates must be documented and susceptible to independent review.

▶ Planing should cover the entire duration of an activity or to some defined planning horizon for projects that have no fixed end point.

▶ Planning preparation must involve performing organizations to a degree sufficient to permit meaningful resource planning.

▶ Plans must be used as the basis for control. Control must include a comparison of actual performance against planned performance, with appropriate action thresholds.

▶ Plans must be kept current, reflecting status and approved changes in scope, schedule, budget, and other important factors.

The following list is a sample outline for a relatively large and complex project's management plan. Use only the parts appropriate for your needs.

1. Work Scope

    a. Purpose of project

    b. Objectives

    c. Deliverables/end products

    d. Key milestones

    e. Requirements

    f.   Key assumptions

2. WBS

3. Organization, responsibilities, and authority

    a. Interorganizational

    b. Intraorganizational

    c. Temporary organization

4. Schedules

5. Budgets and cost estimate basis

6. Resource management

    a. Labor

    b. Equipment

    c. Materials and supplies

    d. Physical facilities

7. Materials management

8. Quality

    a. Codes, standards (e.g., ISO 9000), and regulations

    b. Authorities

    c. Procedures

    d. Records

9. Safety

10. Security

11. Management control

12. Reporting requirements

    a. Management reports

    b. Technical reports

    c. Public information

13. Configuration management

14. Change control process

15. System engineering

16. Environmental protection

17. Data quality

18. Operational definitions

19. Appendixes

   a. Technical data

   b. Support exhibits

   c. Work package documentation

   d. Procedures

   e. Change control board charter

Because the project plan on a large project can get quite large, consider keeping it in an electronic version, available to all team members. Always keep the up-to-date project schedule printed in hard copy and readily available throughout the project work areas.

5.10 Change management

Change management ensures that only changes approved by the project manager are implemented on the project. The most important function of change control is to ensure that everyone working on the project is working to the same plan, including the same scope of work and detailed project requirements. Other functions of change control include:

▶ Making sure people work on only approved changes;

▶ Assessing the impact of changes on cost or schedule before deciding to implement them;

▶ Enabling billing of the customer for customer-directed changes;

▶ Providing a record of changes;

▶ Providing traceability to the original project baseline.

For larger projects, change control may be part of your project quality system. For smaller projects, it may be a memo from the project manager approving the change and identifying the latest version of the specifications and project plan.

5.11 Project closure

Project plans often neglect closure of the project. Project closure includes dealing with the entire project administrative, facility, and personnel issues as the project is finally completed. It usually involves final billing, disposition of project records, and closing the project office. For organizations that perform multiple projects, it also should include a "lessons learned" assessment, to improve processes on future projects.

5.12 Summary

This chapter provided the process and tools necessary to create an effective project work plan.

▶ The project charter is a necessary precursor to a successful project plan that effectively meets all project stakeholder requirements.

▶ The work breakdown structure logically defines the general project work scope and provides the framework for responsibility assignment.

▶ The stakeholder endorsed project work plan defines the scope, schedule, responsibilities, and budget for the project.

▶ Project networks should be as simple as possible to perform the project.

▶ The project plan requires a correct, resource-loaded logic network to develop the schedule.

▶ Dates are outputs from the logic network, not inputs.

▶ If cost is important to your projects, include a cost buffer in the cost estimate.

▶ Request task duration estimates initially, just as you have in the past, and then go back to request average times for the critical chain plan.

▶ Most projects require a change control process.

▶ All project plans should consider project closure as part of the plan.

Adjust the degree of detail you put into the project plan and the degree of formality you put into the project documents to match the stakeholder and team member needs. In general, larger, longer, and government projects require more detail and more formality. Less experienced teams may also require more documentation and training.

References

[1]   Duncan, W. R., et al., *A Guide to the Project Management Body of Knowledge*, Upper Darby, PA: Project Management Institute, 1996.

[2]   CH2MHILL, *Project Delivery System: A System and Process for Benchmark Performance*, CH2MHILL, Denver, CO, 1996.

[3]   Goldratt, E. M., *It's Not Luck*, Great Barrington MA: North River Press, 1994.

[4]   Dettmer, H. W., *Goldratt's Theory of Constraints*, Milwaukee, WI: ASQC, 1997.

[5]   Kerzner, H., *Project Management: A Systems Approach to Planning, Scheduling, and Controlling*, 4th ed., New York: Van Nostrand, 1992.

[6]   Kerzner, H., *Project Management: A Systems Approach to Planning, Scheduling, and Controlling*, 6th ed., New York: John Wiley & Sons, 1998, Table 14-12, p. 745.

[7]   Kiley, M. D., and A. Marques, *1997 National Construction Estimator*, Craftsman Book Company, 1997.

[8]   Vigder, M. R., and A. W. Kark, "Software Cost Estimation and Control," NRC-CNRC (National Research Council Canada), NRC No. 37166, Feb. 1994.

Contents

Developing the (single-project) critical chain plan

This chapter first presents the overall process to create the single-project critical chain plan and then takes you through examples and exercises to practice the ideas. It is important to understand the ideas before you begin to use computer schedule aids to develop critical chain project plans.

6.1   The process

The basic steps of the process to create a single critical chain project schedule follow. We have written the procedure assuming that you are not using a critical chain computer scheduling tool such as ProChain™ or Concerto™. Working some networks by hand helps you to understand the problem the computer is solving. That will aid greatly when you are diagnosing unexpected results.

1. Identify the critical chain.

    a. Lay out the late-finish network of tasks. The tasks must identify the time duration estimate (50-50 time) and primary resource requirements. (For tasks with multiple resources, identify the primary resource you believe will be a constraint. If there are several constraint resources, break the task up for each primary resource.)

    b. If you do not have resource contention in your project, go to step 1(f).

    c. Identify the contention you will resolve first. That should be the contention nearest project completion or the one that shows the most conflict. If several contentions show about the same amount of potential conflict, choose the first one you come to working backward from the end of the schedule.

    d. Remove resource contention by resequencing tasks earlier in time. (Do not worry about creating new conflicts with this step; you will resolve those in sequence.)

    e. Return to the end of the schedule and follow step 1(d) for the next resource. As you resolve conflicts for the next resource, you must maintain the lack of the conflict for the resources you resolved earlier. Repeat until all identified resource types are resolved.

    f. Identify the critical chain as the longest chain of dependent events.

2. Exploit the critical chain.

    a. Review your plan to determine if there are obvious ways that resequencing can shorten the overall project duration. If so, do it. Do not spend a lot of time trial and error testing various solutions. You will usually get a good enough solution on your first or second try.

    b. Add the project buffer to the end of the critical chain.

    c. Add resource buffers to the critical chain.

3. Subordinate the other tasks, paths, and resources to the critical chain.

   a. Protect the critical chain by adding CCFBs to all chains that feed the critical chain. Size the buffers using the longest preceding path. (Note: All noncritical chains feed the critical chain to complete the project. If chains go directly to the project buffer, they also need CCFBs.)

   b. Resolve any resource contentions created by adding feeding buffers through resequencing tasks earlier in time.

   c. Move earlier in time any dependent tasks preceding those moved.

4. Elevate (shorten) the lead time of the project by using added resources for certain windows of time to break contention.

5. Go back to step 1, identify the critical chain. Do not allow inertia to become the constraint.

A critical chain plan schedules (i.e., assigns dates) only to the start of the chains and the completion of the project. Avoid publishing and discussing individual task start and complete dates—they are meaningless. For that reason, you may want to consider talking about the critical chain plan, rather than the critical chain schedule.

6.2 The "good enough" concept

"Good enough" is an important idea in developing critical chain project plans. No proven effective algorithm exists for resource leveling to ensure an optimum schedule. The procedure for developing the critical chain plan ensures that the plan you build will be good enough. That means the overall length of the schedule will close to the shortest or optimum schedule path. *Close* means within a small part (25% or less) of the project buffer. Because reality will change many assumptions, and we cannot explicitly predict the results of statistical fluctuations, that is "good enough."

6.3   Examples and practice

6.3.1   Small example

This section presents a small example to work into a critical chain. Figure 6.1 illustrates the plan in a conventional critical path display, with the early-start schedule. The first number on each bar is the WBS task identification. The number in parentheses is the task duration, in days. Note that task 3 depends on the completion of tasks 1.2 and 2.2.

Following the procedure, in Figure 6.2 we first cut the task times to the 50-50 estimate and push all the tasks to the latest time possible, considering the network dependency.

Next, in Figure 6.3, we add the project buffer and the feeding buffer. Because all tasks use the same resource, we do not need to add resource buffers to this project.

Now consider the same small project with resource contention. Figure 6.4 shows the unscaled network of tasks with a PERT chart representation of the project. The network shows the different resources as colors. You can think of the colors as different skills, e.g., engineers, musicians, or equipment operators.

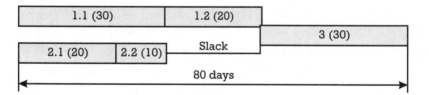

Figure 6.1   A simple project illustrates a normal early-start schedule.

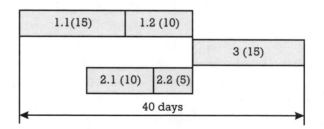

Figure 6.2   The first step to create the critical chain reduces the task times and organizes tasks to a late-finish schedule.

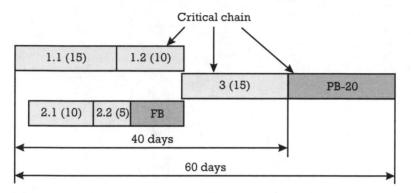

Figure 6.3 With no resource contention, just add buffers.

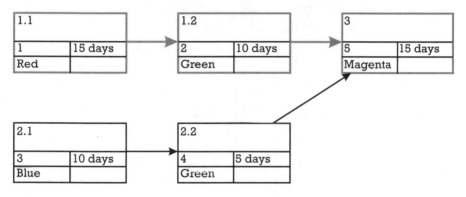

Figure 6.4 The same projects with specific resource assignments (colors).

We lay the network out, with all the tasks pushed as late as possible. In this case, all we have to do is add a start-as-late-as-possible constraint to task 2.1 (Figure 6.5).

Next, remove the resource conflict, working backward from the end of the project.

Figure 6.6 illustrates the two ways we could resolve the green resource contention. Note that each schedule shows a new dependency for the resource constraint. Which resolution choice is better? You may initially think that the lower choice is better, because it is a shorter schedule. With the lower choice, the two chains are the same length, so we could choose either one as the critical chain. Add the project buffer and CCFB to each option and see what happens (Figure 6.7).

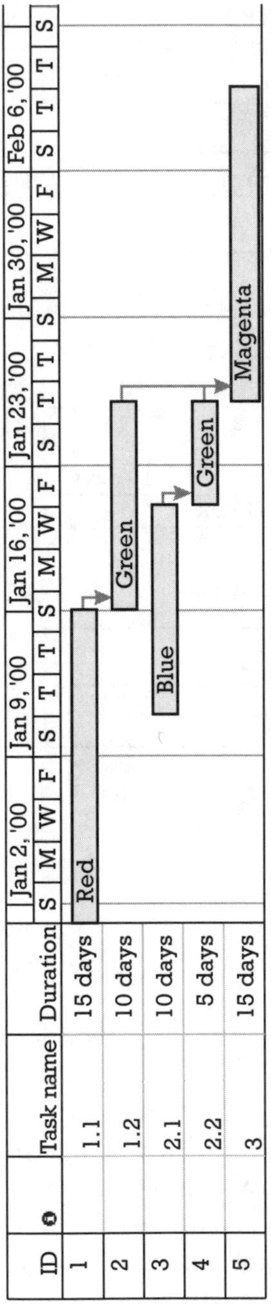

ID	❂	Task name	Duration	Jan 2, '00			Jan 9, '00				Jan 16, '00				Jan 23, '00			Jan 30, '00			Feb 6, '00		
				S	M	W	F	S	T	T	S	M	W	F	S	T	T	S	M	W	F	S	T	T	S
1		1.1	15 days	Red																					
2		1.2	10 days								Green															
3		2.1	10 days					Blue																		
4		2.2	5 days												Green											
5		3	15 days														Magenta									

Figure 6.5   The first step pushes tasks to the late finish.

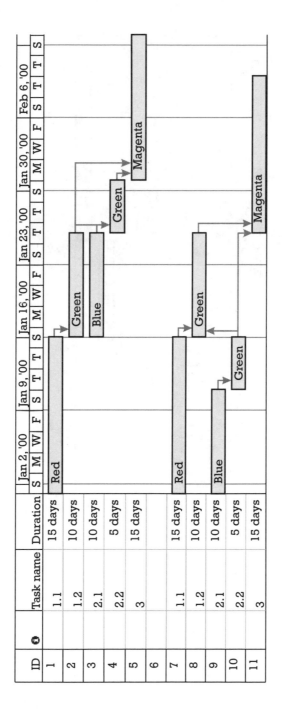

Figure 6.6 Alternative ways to resolve resource conflict.

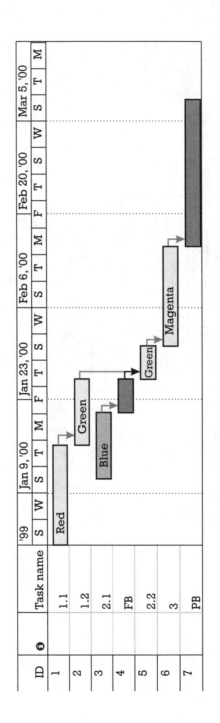

Figure 6.7 The first choice results in a planned completion date of March 7, 2000.

Figure 6.8 illustrates the critical chain plan for the first choice of removing conflict for the green resource. The critical chain comprises all the project tasks except task 2.1. The buffers are sized as 50% of the feeding chains. (In a case like this, we recommend a modification of that later for the feeding buffer.)

The two choices of critical chain from the first resource resolution option both lead to the same overall length of schedule. Both also create the situation in which the noncritical chain is longer than the critical chain after the addition of the feeding buffer. That is all right; we just have the extra lead time for the noncritical chain paths. That often happens when a schedule starts with two or more paths of nearly equal length. If you are not comfortable with this method, you may move the excess feeding buffer (i.e., the amount that would push the non-critical chain earlier than the critical chain) to just in front of the project buffer. When you use buffer management, act as if all the feeding buffers were together at the end of the feeding chain.

Any of those paths is suitable for the plan because the differences are small compared to the project buffer. It often works out a little better to resolve contention by moving the longer of the two or more tasks backward in time. That tends to keep the critical chain the longest chain, thereby increasing the project buffer and adding to the immunity of your project.

Lay out the exercise as a critical chain plan, with all the appropriate buffers. In Figure 6.9, the first line in each box represents the task number. The second line represents the resource, by color. The third line represents the (already reduced) task time, in days. (See the last question in Section 6.10 for the approximate length of the schedule you should have obtained. Note that a good enough schedule is within a small part of the project buffer of the schedule given in the back of the book.)

6.3.2 Large example

Figure 6.10 presents the task network for the large example. The top line of each box is the identifier for the task; the color relates to a specific resource; and the number at the bottom of each box represents the task duration, in days. The task durations have already been cut by 50%. Lay out the critical chain plan for this project.

The first step is to lay out the late-finish sequence of tasks. To do that, you have to convert the network plan to tasks that give the planned duration. Figure 6.11 shows the project laid out with Microsoft Project 98

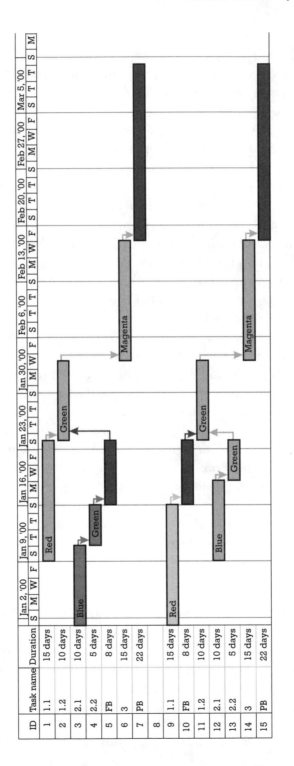

Figure 6.8 The second resource resolution choice leads to the same lead time for the two choices of critical chain and completion on March 10, 2000.

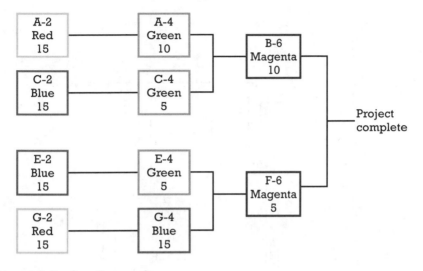

Figure 6.9   Small exercise.

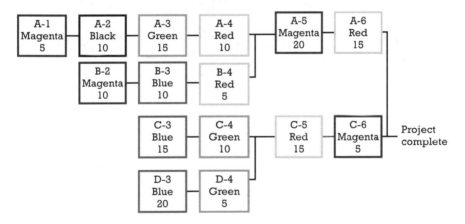

Figure 6.10   Large example. *(Task times already reduced.)*

using the as-late-as-possible constraint. The calendar is set to work through the weekends.

Work from the end of the project forward in time to remove the resource contention. The first conflict is between two tasks using red, A-6 and C-5. Because the A path is the critical path, it usually makes sense to give it the resource first, which means moving the A path forward in time, and the B path along with it, as illustrated on the following page. Figure 6.12 shows the plan with the contention removed.

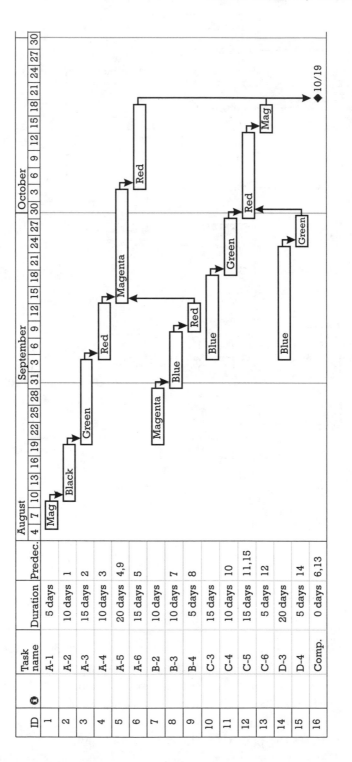

Figure 6.11 Large example time-scaled logic illustrates resource contention.

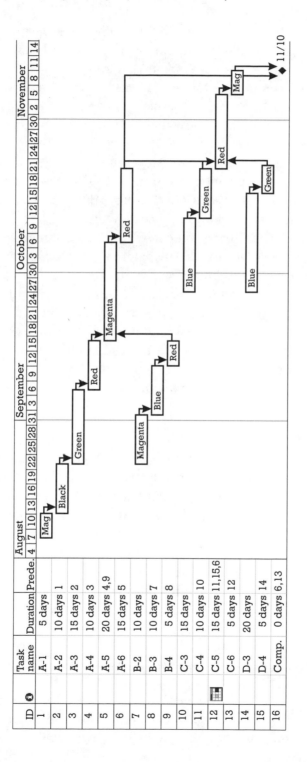

Figure 6.12 Resolving the initial (A6-C5) red-resource contention.

The next conflict is green (C4 and D4), followed by blue (C3 and D3). Because those paths and tasks will not be on the critical chain, and both are the same length, the order is somewhat arbitrary. You then have to remove the red conflict (A-4 and B-4), and so on. This first round of conflict removal leads to the project plan shown in Figure 6.13.

You can now select the critical chain and size the project buffer. The critical chain comprises tasks A-1, A-2, A-3, A-4, B-4, A-5, A-6, C-5, and C-6, which add up to 100 days. We should set the project buffer size as 50 days, for an overall planned lead time of 150 days for the project.

Next, you have to add the CCFBs. The tasks that feed the critical chain are B-4, C-4, and D-4. Working backward, we can add the feeding buffer for D-4 (22 days). That creates some new conflicts. C-4 requires a 22-day feeding buffer also. We have to do D-4 before we do C-4, so if D-4 is late, it will show up in the buffer from C-4.

The final step is to add the resource buffers. (We could have added them immediately after identifying the critical chain but did not, simply to keep the chart clearer until we were done.) The final plan (Figure 6.14) shows the resource buffers as boxed Rs. The buffer goes with the task immediately above it.

This completes the feasible and immune plan.

6.3.3   Large exercise

Lay out the exercise in Figure 6.15 as a critical chain plan, with all the appropriate buffers. As in the small exercise, the first line in each box represents the task number. The second line represents the resource, by color. The third line represents the (already reduced) task time, in days. (See the last questions in Section 6.10 for the approximate length of the schedule you should have obtained.) Note that a good enough schedule is within a small part of the project buffer of the schedule given in Section 6.10. You can find a picture of the completed plan on Advanced-projects.com.

6.4   Buffer and threshold sizing

Buffer sizing determines the overall duration of your project and the degree of overall contingency included in the plan. The buffer thresholds for action determine the frequency with which you will act. We usually

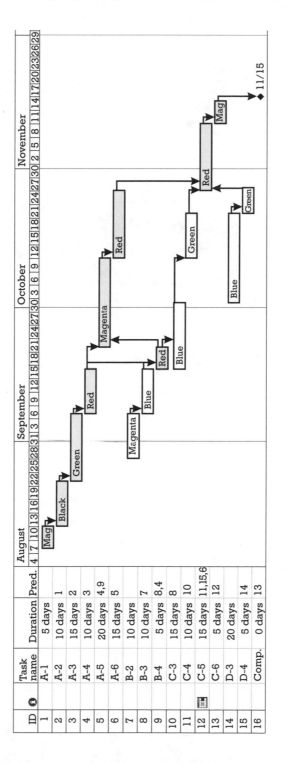

Figure 6.13 Resolving other resource contentions.

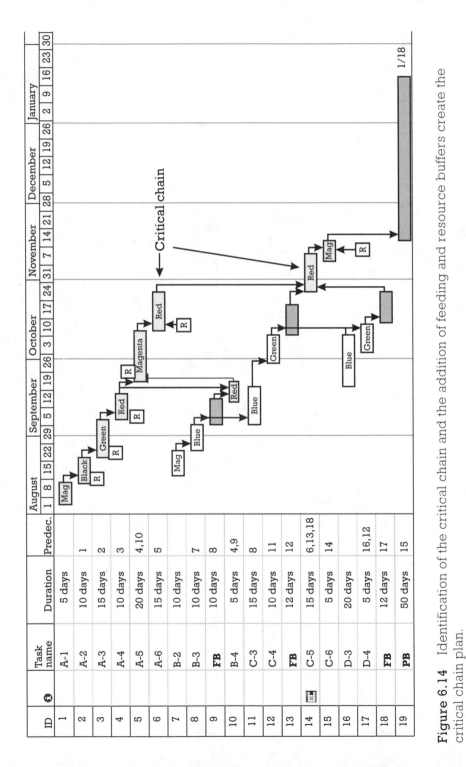

Figure 6.14 Identification of the critical chain and the addition of feeding and resource buffers create the critical chain plan.

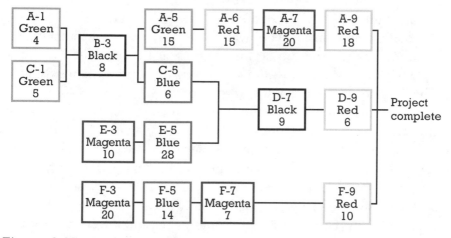

Figure 6.15   Large exercise.

set buffer thresholds as a percentage of the buffer, so the buffer size influences the actual sensitivity of the buffer triggers.

6.4.1   Statistical background

Recommendations on buffer sizing use statistics to develop relatively simple rules with a supporting theoretical basis. Dr. Goldratt recommends sizing the project buffer and feeding buffers to one-half the buffered path task length. That is, do not include gaps in the chain when you are sizing buffers. The buffers are there to protect the project from uncertainty in performing the tasks on the chain.

Goldratt's method considers the statistical rule governing the addition of uncertainties that are independent events. The statistical rule says that the uncertainty of the sum of the events is much less than the sum of the uncertainty for each event. That is sensible, because you should expect some variations to be positive and some to be negative. Consider Dr. Goldratt's recommendation in context with his recommendation to simply cut activity times in half. Mathematical justification of his recommendation requires several additional assumptions, some of which we highlight here. His recommendation usually will lead to larger buffers than the method described next, a reasonable thing to do when you are beginning to deploy critical chain.

The spread in a distribution is proportional to the standard deviation, σ or sigma. The spread of the distribution representing the sum (in our case, the buffer) equals the square root of the sum of the squares of the individual distributions. (Do not worry if you are not a statistics buff and cannot follow this. You can do fine with critical chain using Goldratt's simple recommendation or simply by following the procedure we give below. You do not have to know this theory to have it work for you.)

If you make a few assumptions, you can come up with a relatively simple way to make use of your knowledge of the variation in estimates to size the project and feeding buffers. Projects usually do not have much information about the actual distribution of the task performance time. (Exceptions might include repetitive projects, such as construction, in which extensive cost data exist.) However, you can usually place bounds on the task time, corresponding to some upper and lower limits of the time it will take. If you assume your estimating method yields about the same meaning for the upper and lower limits on most of the project tasks, you can then say that the difference between the upper and lower limits, D, is some multiple of the standard deviation. You may not know if it represents two or six standard deviations; you are only assuming that whatever it is, it is about the same for all the tasks you estimate with the same method. Then, without even having to define the limits precisely, you can size the buffer to protect the whole chain of tasks to the same degree we previously were protecting each activity. You take the square root of the sum of the squares of the Ds. The result is always less than adding the Ds.

For example, consider a chain of four tasks, each two weeks long. Two weeks is our standard low-risk estimate. One week is our 50-50 estimate. So D equals 1.

The critical path chain is, therefore, eight weeks. The critical chain tasks add up to four weeks. Because D equals 1, D squared also equals 1. The sum of D squared is then 4, and the square root of 4 is 2. Adding the two-week buffer to the four-week task chain gives a project duration estimate of six, compared to eight for the critical path. In this case, the square root of the sum of the D-squared method gives the same result as Dr. Goldratt's simplified method. That always happens for four equal-length tasks, where D is half the task duration, that is, not very often.

6.4.2 Project buffer size

Size the project buffer using the square root of the sum of the squares method. Determine the D value for each task as the difference between the initial task duration estimate and the reduced estimate. The following guidelines will help ensure an effective buffer:

▶ Seek to have at least 10 activities on the critical chain. *Reason:* The more activities in the critical chain, the more effective the sum of the squares and central limit theorem.

▶ Do not allow any one activity to be more than 20% of the critical chain. *Reason:* The uncertainty of one large activity will dominate the chain, leaving little possibility for the other tasks in the chain to make up overruns on the dominant task.

▶ Do not allow the project buffer to be less than 25% of the critical chain. *Reason:* Chains with many tasks of uniform length may calculate a relatively small buffer, providing inadequate protection.

6.4.3 Feeding buffer size

Size the feeding buffers using the square root of the sum of the squares method. Determine the *D* value for each task as the difference between the initial task duration estimate and the reduced estimate.

If there are fewer than four tasks in the feeding chain, make sure the feeding buffer is at least equal to the longest activity in the feeding chain.

6.4.4 Buffer trigger points

We set the buffer trigger points to plan for management control action and to initiate the action. Both trigger points must be set to minimize false signals and to ensure that action is taken when needed. It does not damage project performance directly to plan for project changes that are not made. Thus, there is less negative impact from too low a threshold for the project plan (yellow) trigger point. You may do significant damage to your project, however, if you set the action (red) trigger too low and take unnecessary control actions. Project changes, which include control actions, will likely cause confusion and delay the project.

We suggest setting the triggers at one-third and two-thirds of the buffer. Because project tasks are not always in a provable state of statistical control, we recommend that you track buffer penetration over time. If you are tracking the buffer over time, you may want to institute some additional control chart triggers, such as four points in a row tending toward the trigger point. Do not make the trigger logic too complex.

Some people suggest that the trigger points should be relative or dynamic. That is, the triggers should require less penetration early in the project. The logic is that early in the project people may be inclined to use up the buffer. That fear, however, most often is baseless. Usually, there is negative buffer penetration early in the project. We suggest you trend the buffers and make decisions as you deem necessary. Be mindful that too many control actions have a negative effect on project performance.

Set the buffer triggers for feeding buffers at the same percentage of buffer penetration as for the project buffer.

6.4.5   Resource buffer size

Size resource buffers to the needs of the resource provider. The size should depend on the quantity of the resource, the length of the resource's usual task, and special considerations such as required training, travel, or other lead time.

For subcontractors, consider making the resource buffer a financial incentive to ensure a lead time. Because profits are a small percentage of revenue, you are often able to greatly increase delivery reliability by doubling the suppliers' profit if they deliver on time, which should cost only a small percentage of the subcontract. A recent public example is that of the contractor who rebuilt an overpass on the Santa Monica Freeway, that was destroyed in an earthquake, and finished over a month early due to a significant reward.

6.5   Cost buffer

Use a cost buffer if your business is sensitive to project cost. Organizations that use throughput accounting and internal projects (e.g., internally funded R&D) may not require a cost buffer.

Size the cost buffer taking into consideration the amount removed from the project plan when the activity durations were reduced. You can use the sum of the squares to size the required buffer, where the Ds are the cost reduction of each task. That will give a project budget, including a cost buffer, that is significantly less than the original budget.

If your organization uses cost and schedule control reporting or project plans to sum up organizational resource demands, add the cost buffer into the project plan. We recommend you put it all into the project buffer.

If you use other means for global resource planning, you can put it in the buffer as a leveled fixed cost. If you use the individual project plans to project resource demand, you must put in a resource distribution representative of the aggregated project. For example, divide the people resources to represent the same percentage in the buffer as they are in the plan.

6.6 Methods to create the plan

People have successfully used a variety of methods to make and control critical chain plans. Initial critical chain projects all used some type of manual method. Keep in mind that we are cautioning against putting too many tasks in a critical chain plan (i.e., a critical chain plan should have no more than a few hundred activities, preferably fewer than 100).

6.6.1 Manual method

The simplest and most commonly used method to manually create a plan is to use the PERT chart format and sticky notes. The procedure follows:

1. Fill out a sticky note for each task, containing the task ID, title, duration (reduced), and controlling resources. (You may want to use color coding to identify the task duration controlling resource.) On the left of the note, indicate the tasks that provide needed input.

2. Lay the notes out on a board or table according to the task logic and following the rough time logic (this is called a time-phased PERT or a time-phased logic diagram).

3. Remove resource contentions.

4. Identify the critical chain.

5. Add sticky notes for the project and feeding buffers.

6. Size the feeding buffers.

7. Calculate the critical chain using a forward pass. Starting with the initial task, write the start times on the lower left of the note and

the completion time (start time plus duration) on the lower right corner.

8. Calculate the feeding paths using a backward pass from where they enter the critical chain.

9. Remove any remaining resource contention and revise the calculation.

10. Identify the locations for the resource buffers.

11. Size the resource buffers.

This process is not difficult for projects with 10 to 50 tasks. It gets harder after that.

You may refine the method by cutting out colored paper bars to represent each task. The length of the bar represents task length, and the bar color represents the task duration controlling resource. That simplifies the resource contention steps and subsequent calculation. It obviously requires a little more upfront preparation. Large projects have used this method successfully with over 500 tasks. Using a magnetic scheduling board is another way to implement the same idea.

6.6.2 Critical path software

You can use critical path software to plan and manage critical chain projects. Most software packages have sufficient options to support you in leveling the resources and using late start on the feeding chains. You always start from the same place: with a project logic containing the reduced task times and resource requirements. You should ensure (when necessary) that you have selected the appropriate options to maintain the fixed task duration that you input and that you have selected options to late-start each path. Sometimes, you can do that globally. Other times, you can put constraints on the first task on each path that causes all the downstream tasks to late-start. (You need to experiment and understand what your software does to those options or constraints during resource leveling.)

Most critical path software provides options for the algorithm to perform resource leveling. You can experiment with them. The critical chain method does not depend on the algorithm you use. It simply requires that the final plan have removed all resource contention within the single project. Usually, you can do resource leveling manually and view the final resource allocations by task.

After initially leveling resources, you must identify the critical chain. We suggest you add links to the plan to cause the resource leveling to stay in place. You can then remove other constraints that your software may have added to implement resource leveling (e.g., some software adds fixed task start-date constraints to implement resource leveling). If you do add logic connections, you then should be able to calculate the schedule and have the critical path equal the critical chain.

Make sure the critical chain you identify really is the constraint of your project. Sometimes an inadvertent logic connection results in tasks on the critical chain that cannot or should not determine the duration of your project. (We call such a connection a mathematical critical path/chain.) Adjust logic or task duration to cause the critical chain to be a legitimate constraint to your project. (Note again that there may be two nearly equal length paths vying for the critical chain. We suggest you choose the one that you feel has higher uncertainty or that makes most use of a potentially capacity-constrained resource.)

The distribution of tasks on the critical chain must provide effective immunity from variation in any one task. There are two simple guidelines for doing that.

1. Make sure the critical chain comprises at least 10 tasks (unless your project is very small).

2. Make sure no single critical chain task comprises more than about 20% of your critical chain or more than 50% of your project buffer.

Next, add the feeding buffers and the project buffer. You add these as tasks, without resource requirements. Remember to tie in the feeding buffers as predecessors to the critical chain task at the point they join the critical chain. Then recheck the resource leveling and make any final adjustments. (Adding the feeding buffers usually requires redoing some amount of resource leveling.)

6.6.3 Critical chain software

Critical chain software automates most or all of the process. Several software packages are currently available, and we understand that most major project management software will be adding the capability over the next year. The most widely available software currently used is ProChain™, an add-on to Microsoft Project. Concerto™ is another currently available product.

6.7 External constraints

Projects may have external constraints, which can influence the project lead time and which are not under the control of the project team. Regulations, inspections, and permits often fall into this category. External constraints may be internal to the company, such as another division that has to provide an essential component.

The five focusing steps provide a method to deal with external constraints. First, you must identify them as constraints (or as potential constraints) and deal with them accordingly. If they are only potential constraints, you can deal with them under project risk management. If you feel that the likelihood for a potential constraint becoming an actual constraint is large, you may want to make sure that it is on the critical chain.

The second step is to exploit the constraint. In the case of regulations and permits, that usually requires providing a high assurance that all submissions to the regulators meet their needs completely. That may require additional resources upfront. You should consider, however, that any delay in the project critical chain should be valued for the burn rate of the entire project or the expected daily return upon completion of the project. You may elect to hire experts in the particular area to help ensure success. There may be portions of the project that can be exempted from the constraint.

The third focusing step subordinates everything else to the constraint. That may require doing additional scope or investing additional management time to ensure good working relationships to any people or agencies that may become an external constraint, as it is not usually under your control.

It is improbable you would elect to elevate an external constraint.

6.8 Reducing planned time (a.k.a. dictated end dates)

Project managers are often asked to accelerate schedules. With CCPM, there may be a tendency to look at the juicy project buffer and suggest that reducing the buffer is a painless way of reducing the planned project lead time. Reducing the project buffer has no impact on project execution time; it only reduces the chances that you will meet your promised lead time and causes excessive buffer triggers. Excessive buffer triggers damage project performance. Therefore, do *not* cut the project buffer.

6.8.1 Acceleration without cost impact (exploit and subordinate to the constraint)

Several sensible methods can reduce project lead time. Preferred options do not increase cost. Two primary options are to get additional resources when resolving contentions caused the lead time to be increased and to look inside the tasks for batching opportunities.

You may need only a short time of an additional resource to make a significant improvement in the project overall lead time. If there is a way to obtain the additional resource, this method can reduce the overall project lead time at no additional cost, since you had to perform the tasks for the same individual durations, that is, you do not change the task work (person-days). You can reduce the project buffer if such a change reduces the length of the critical chain.

Batching occurs when a task includes more than one physical output. For example, a task may include making a number of certain parts used in the final assembly. The parts may be identical or different and are not limited to hardware. They might include different technical products, such as drawings, parts lists, or reports, or even different people, such as hiring people to staff one shift at a time.

The successor task may be able to start when the first of the predecessor outputs is available. In that case, you can break up the task into smaller pieces to better show the real workflow. Your plan can also show that type of relationship as a task start-to-start dependence, with a lag. Alternatively, you can show it as a finish-to-finish task logic. Whichever way you choose, your management process should ensure that performers understand and focus Roadrunner performance (i.e., start the task as soon as input is available, focus on it 100% until done and pass on your result) on each individual task output. They must keep the sequence needed to realize the assumptions made in your plan.

If batching involves a significant number of parts, you may want to invoke a supplemental method to track and control the parts through the repetitive process. The critical chain plan would show the process as a single activity, for example, "Process 37 parts." One effective method uses the line-of-balance method, combining features of operational process control with project management. The line-of-balance method plans the time for each part to traverse the process flow, creating an expected number of parts through each step at a given time (the line of balance). Tracking compares the actual parts through each process step to the line of balance.

6.8.2 Acceleration with increased raw material cost (elevate the constraint)

You can also reduce project time by exercising higher-cost alternatives. For example, you can use overtime or hire additional temporary resources (which usually cost more). You may be able to purchase components with a higher cost but a shorter lead time. You may be able to use higher premiums for early subcontract delivery.

TOC suggests that considerations of increased cost compare the additional operating expense to the impact on project throughput. The throughput of project acceleration (per day) is the value of the whole project (per day). Compare the cost of increased raw material cost to the throughput increase from the acceleration. If the throughput increase exceeds the cost increase, you should elevate the constraint.

6.9 Enterprisewide resource planning

You can use enterprisewide scheduling tools to identify to the resource managers the anticipated window and duration of tasks they have to support. Resource managers must buy in to understand that the dates are not meaningful; we focus on windows of performance and task duration estimates. The resource managers can then assess if their long-term aggregate demand requires more or fewer resources and allocate resources based on the criteria.

6.10 Frequently asked questions

Sometimes, abnormal things seem to happen, and questions arise. This section addresses those situations and questions.

▶ *After we add the feeding buffers, noncritical chains start earlier than the critical chain. Why?*

That can happen and should not be a cause for concern. Start the project with the noncritical chain. Be sure to use a resource buffer on the first task on the critical chain. An alternative method relocates the excess feeding buffer (i.e., that which pushes the feeding path before the critical chain) just ahead of the project buffer.

▶ *When we add the feeding buffer to a noncritical chain with a critical chain task as a logical predecessor, it pushes the critical chain task back, creating a gap in the critical chain.*

If the change required is a small percentage of the project buffer, make the critical chain feeding buffer a little smaller than the buffer sizing indicates. For other cases, consider where the noncritical chain feeds the critical chain and the relative variability of the two chains. You may also relocate the excess feeding buffer to just before the project buffer. Remember, we are subordinating everything else to the critical chain. In general, gaps in the critical chain should happen only because of a company constraint resource. Gaps in the plan do not mean you should have a gap in performance.

▶ *Why do we not connect the other chains by their resource and path dependencies?*

It is not necessary, and attempts to add that level of detail do not improve project performance. Fluctuations will occur, so attempting to control every dependent chain is not possible. The CCFBs and the use of buffer management provide the necessary and sufficient control.

▶ *Our schedules have thousands of tasks, so there is no way to plan the project without an effective computer program.*

Question your assumptions. Is it logical to assume that increasingly detailed complexity in a project plan will make the project more likely to succeed? Can a project manager really manage thousands of tasks? Does having thousands of tasks improve the ability to manage, or does the system become so complex that it can never have meaningful accurate data? How important can it be to have tasks as small as a fraction of 1% of the total project? (With 100 tasks, the average task size is already down to 1% of the overall project. Who can estimate that well?)

We recommend that you confine project schedules to a few hundred tasks, at most. If major subassemblies (e.g., an aircraft engine) require schedules of their own at a lower level, use that approach. (We are working the project management system here, not detailed designs or bills of materials.)

Experience demonstrates that the more detailed tasks there are in the schedule, the more often the schedule has to be revised and the greater the probability of error. That leads to long turn-around times for schedule updates and the loss of control.

▶ *We are halfway through the project and have not penetrated the project buffer. Can we cut the project buffer in half?*

Cutting the project buffer does not reduce the project actual performance time. It reduces the chance that the project will deliver substantially early. Your project buffer status gives you dynamic predictions of project completion time. There is no reason internal to the project to reduce the project buffer.

If external needs require you to reduce the project buffer, you can replan the project at any time. Remember, the project buffer protects the whole project. All noncritical chains merge to the critical chain before the project is complete. Before you reduce the project buffer, check all CCFBs to make sure that the unused feeding buffer length is at least 50% of each feeding chain uncompleted path length. If the feeding buffers are all intact by that amount, there is no problem with reducing the project buffer to 50% of the remaining length of the critical chain. In essence, you are starting a new project at the time of the update.

▶ *We have tasks in our project plan over which we have no control. What should we do?*

Regulator or client review of project outputs often creates that situation. You can control what you give them and when you give it to them, but you cannot (directly) control their work processes. In this case, working with your stakeholders, as described in Chapter 1, will provide great benefit. You can influence how long their review takes and limit potential rework by using the effort necessary to ensure that you understand their requirements and produce a quality product for their review. If you are a significant part of their workload, you can help their focus by staggering your submissions to help them avoid multitasking. Other unique situations demand unique solutions. In those instances, use the five-step focusing process (see Section 2.3.3).

▶ *Our management/client has specific intermediate milestones they want us to schedule a date for and meet. What do we do?*

This can occur for a number of reasons, including coordination with other parts of a larger project. We know of cases in which project payment is tied to satisfactory completion of milestones.

If satisfying milestones creates throughput for your company, we recommend planning milestone accomplishment as a project of its own. You can then use the multiproject method to link the projects.

If satisfying the milestones is simply a tracking tool, we suggest you first try to convince management or the client that buffer reports are actually a better tool. Failing that, we suggest you protect the milestones with a milestone buffer. Size the milestone buffer as a project buffer, but do not use it for project control.

▶ *Our client does not want our result early, because we are a subassembly to their project, and they do not want to have to store our input. What do we do?*

Use the critical chain process to schedule the start of your activity chains to satisfy the client needs. Usually, that will mean you can delay some activity starts.

▶ *What are the answers to the two exercises in Section 6.3?*

There are multiple satisfactory solutions to each exercise. If your results come within about 15% of the project buffer to the total lead times given below, they are good enough.

	Critical Chain Length	Project Buffer	Total Project Lead Time
Small exercise	50	25	82
Large exercise	107	47	154

6.11 Summary

This chapter described how to create a critical chain plan for a single project. The steps up through creating a logic diagram with low risk duration estimates do not change from the reference PMBOK approach. The critical chain steps are as follows.

▶ Estimating task duration for the critical chain plan is often one of the most challenging implementation problems. Be sure you create a plan with conventional duration estimates before you ask for 50-50 estimates.

▶ The critical chain plan process is well defined and easy to use to create a "good enough" plan.

▶ Project and feeding buffer sizing and trigger points determine the degree of project protection and frequency of control action.

▶ Size resource buffers to send effective signals to resources on impending need.

▶ The (optional) cost buffer provides aggregated cost protection in the same way that the project buffer protects the schedule.

▶ You can use alternative methods to create and track the critical chain plan, ranging from manual methods to critical chain software.

▶ The TOC five focusing steps (identify, exploit, subordinate, elevate, avoid inertia) provide a framework for resolving environment and project specific issues.

Constructing a critical chain plan is a relatively small addition to the work necessary to construct an effective critical path plan. It may be less work and create a more accurate plan, if you significantly reduce the number of activities in your plan. The extra investment is well worth the gain.

CHAPTER

7

Contents

Developing the enterprise multiproject critical chain plan

7.1 Identifying the multiproject constraint

The critical chain is the constraint for a single project. What is the constraint of an enterprise that performs multiple projects? How do you put the critical chains of multiple projects together in a way that identifies the constraint of the enterprise to produce projects that meet the three necessary conditions and do it in a way that allows focus on increasing the project throughput of the enterprise? What is it that constrains the enterprise from completing more projects or completing the existing projects more quickly?

Consider a more familiar reference environment with which most people are familiar: mowing a lawn. Consider the amount of grass cut as the counterpart to completed projects. What happens when the grass is too long or when you try to push the lawnmower too fast? It bogs down and often stalls.

The same thing happens if too many projects are pushed into a multi-project environment without considering the capability of the constraint to perform the projects. If you push too many projects into the system, it will bog down and stall. People will work hard, but projects will take a long time to complete (the engine is stalled much of the time), and a lot of management effort goes into restarting the engine and cleaning out the debris. It will seem as if there are never enough of the key resources necessary to complete the projects.

With the lawnmower, you use the feedback from the system to adjust the rate of processing. You listen and slow down the lawnmower as the engine begins to slow down. Or you raise the cutting height, so you match the processing rate to the feed of the work.

Figure 7.1 illustrates a critical path multiproject scenario. The colors in the bars represent resources. Using conventional low-risk activity estimates and considering three-project multitasking, each activity duration is 90 days. In most organizations, the managers of the three projects would rarely work together. Each would work with the managers of the resources to try to get the resources they needed. In this worst-case example, all the resource needs overlap. If there is only one of each resource, each project has to schedule assuming one-third of the resources time to work on its project. That situation is called the fractional head count.

I have made a point of asking groups of project managers (many of whom belong to the Project Management Institute, including certi-fied project management professionals), "How many of you routinely resource-level your project plans?" (Resource loading means identifying the resources needed for each task; resource leveling is removing the con-flicts in which demand exceeds supply.) My unofficial survey indicates that only about 5% of project managers routinely resource-level their plans. In other words, the situation is usually worse than I assumed above: They do not even know where the overlaps occur. I then ask them, "Why not?" Most need some prodding, but usually the answer is one of two things: (1) It is not worth it, because management will change every-thing anyhow, or (2) it makes the schedule too long. Finally, I ask how many of them have infinite resources at their disposal. So far, none has infinite resources.

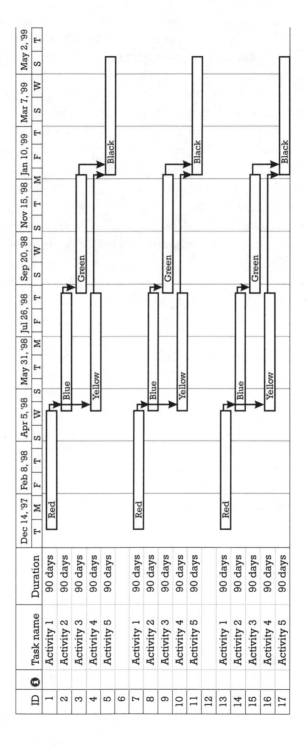

Figure 7.1 Three projects in a multiproject environment.

Some companies do check resource availability across all projects. They then argue to increase resources. That is moving to the elevate stage of TOC, before completing the identify, exploit, and subordinate steps—a very expensive strategy.

Considering that and the Figure 7.1 project, assuming these projects are all the same, the resources have to be divided among the three projects; even if you have only one resource of each type. Thus, either the project plans assume this multitasking, or the projects are not going to complete on time due to the necessity for multitasking. Evidently, one of the resources is the capacity constraint of the system.

You first have to identify the company capacity constraint resource. That is most often a certain type of person, but it may be a physical or even a policy constraint. The company constraint resource becomes the drum for scheduling multiple projects. The terminology comes from Dr. Goldratt's production methodology, in which the drum sets the beat for the entire factory. In our example, the drum set the beat for all the company projects. Think of the drummer on a galleon. What happens if even one rower gets out of beat?

The project system becomes a pull system because the drum schedule determines the sequencing of projects. You pull projects forward in time if the drum completes project work early. You delay subsequent projects when the drum is late. For that reason, projects in a multiproject environment also require buffers to protect the drum, to ensure that they never starve the capacity constraint for work. You also must schedule the projects to ensure that they are ready to use the drum resource, should it become available early.

Figure 7.2 illustrates the CCPM method. Compared to the previous critical path case, you reduce each activity time to 15 days to eliminate the three-times multitasking and to use 50% probable duration estimates. You identify the resource supplying activities 2 and 3 as the capacity constraint resource. You exploit the resource by synchronizing the projects using that resource as the drum. You subordinate to the resource by adding capacity buffers between the projects. The capacity buffers ensure that the capacity constraint resource is available for the subsequent project.

Figure 7.2 shows the CCPM plan completing the three projects (including the project buffer) near the end of August 1998. It shows the first two projects completing even earlier. Compare that to the critical chain multiproject plans of Figure 7.1, all of which are scheduled to

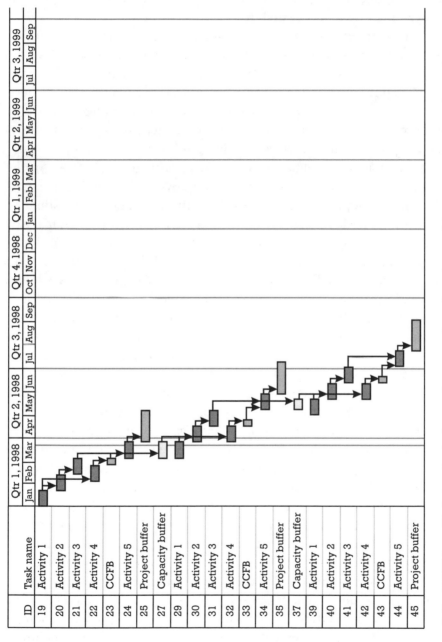

Figure 7.2 CCPM multiple project plan reduces project duration and increases project throughput.

complete in May of 1999. Based on what you have learned for single projects, you can expect the CCPM projects to be early. Based on global project experience, you should expect the critical path projects to be late, for even these extended schedules.

Note that synchronizing the projects this way eliminates resource contention for all resources, not just the drum resource. That happens in the example because the projects are identical. While most multiproject environments do not have identical projects, synchronizing projects to the drum usually eliminates some, if not all, resource contention. Resource manager prioritization of resources according to the penetration of project buffers resolves remaining resource contentions.

This is a major simplification compared to attempts to micromanage a whole enterprise, which never work. By now, you should understand why this is a hopeless exercise. All the activity durations are estimates. None of the activities should take the exact amount of time planned. Any schedule produced for all resources across all projects is fiction. It is only one possibility out of millions of possible combinations of project status and resource availability. Instead, the critical chain process uses buffer management to dynamically allocate resources. CCPM allows for such variation with the resource buffers and feeding buffers within each project. This process also includes the ability to absorb the natural variation in the buffers. It is a real-world control system.

TOC leads to an understanding that all resources other than the constraint must have excess capacity. Those upstream of the constraint resource must have excess capacity to ensure that the constraint resource is never starved for work, which would waste its capacity. In a project, that means we have to buffer to ensure that we provide the constraint resource with the input it needs. Resources downstream of the constraint must have more capacity than the constraint to deal with fluctuations in their own output and that of resources between themselves and the constraint resource. They must ensure that they always deliver the constraint resource-processing rate to the completion of the project(s). In a project, that is the concern of the project, not of the constraint resource.

While projects theoretically can have resource demands in any order, there tends to be a similarity in the order within a company, based on the type of projects they operate. For example, many projects will have a design phase, procurement phase, construction phase, and initial operation phase. Thus, the sequence of demands on resources tends to be similar, although the usage may vary substantially from project to project. The general idea carried over from manufacturing is that the further a resource is from the constraint resource in the plan sequence, the more excess capacity and/or the larger buffer it needs to not affect the overall lead time.

7.2 Exploiting the multiproject constraint

The constraint resource becomes the drum for the company projects (like the drummer on the ancient galleons setting the pace for the rowers). Therefore, the procedure to exploit this resource is as follows:

1. Identify the company constraint resource. The company constraint resource should be the resource that determines the greatest amount of critical chain duration on your projects. It usually will be apparent as the resource that is frequently in short supply and is often called on to use overtime. If several resources exhibit the same behavior, select one based on the unique contribution of your company. Otherwise, select the one usually demanded nearest the beginning of a project.

2. Exploit the company constraint resource.

   a. Prepare the critical chain schedule for each project independently.

   b. Determine the project priority for access to the constraint resource.

   c. Create the constraint resource multiproject schedule: the drum schedule. Collect the constraint demands for each project and resolve contentions among the projects to maximize company throughput. In other words, complete most projects early.

3. Subordinate the individual project schedules.

   a. Schedule each project to start based on the constraint resource schedule.

b. Designate the critical chain as the chain from the first use of the constraint resource to the end of the project.

c. Insert capacity constraint buffers (CCBs) between the individual project schedules, ahead of the scheduled use of the constraint resource. That protects the drum (constraint) schedule by ensuring the input is ready for it.

d. If insertion of the CCBs influences the constraint resource schedule, resolve contentions.

e. Insert drum buffers in each project to ensure that the constraint resource will not be starved for work. Place them immediately preceding the use of the constraint resource in the project.

4. Elevate the capacity of the constraint resource.

5. Go back to step 2 and do not let inertia become the constraint.

Section 7.3 describes the features of this process.

7.3 Features of multiproject critical chains

7.3.1 Project priority

You must prioritize all ongoing projects before you create the drum schedule. The priority is for one purpose: to set the priority for use of the drum resource. Your method for setting the priority may consider a number of factors, but the primary factor from TOC is to prioritize to maximize the company throughput per use of the constraint. If you have a direct measure of project throughput, you can actually use that ratio to set the priority, that is, divide the project throughput (usually in dollars) by the drum resource demand (usually in person-hours or person-days).

Legitimate reasons for other considerations in setting the project priority should consider the company goal. For example, it may be advantageous to give higher priority to your best customers, considering your need to make money in the future.

7.3.2   Selecting the drum resource

The drum resource must be shared across all projects you consider part of the multiproject environment. That is the definition of a multiproject environment. Larger companies may have several independent project groupings that share resources within the group, but not across groups. Only in this case should you have multiple drums.

Resources often appear as constraints. The company capacity constraint sometimes may seem to float. The basic TOC makes it unlikely that there is in fact more than one constraint (unless you have an unstable system!). Statistical fluctuations can make temporary capacity constraints. For example, suppose a number of projects happen to demand a particular resource at one time, exceeding the resource capability. That is a statistical occurrence, and you should expect it to happen. It does not make the resource a company capacity constraint. It does mean the project plan and control system has to handle it, even if only through the individual buffers already added. There is also some flexibility in resource supply, for example:

- ▶ Using overtime or asking people to defer time off;
- ▶ Segmenting the work to ensure that you are properly exploiting the potential constraint;
- ▶ Subordinating other work that does not produce immediate throughput.

However, many companies have a chronic resource constraint: the department that is always on overtime or the one that always seems late. Presumably, that department has been permitted to occupy that position because of some policy or other reason that prohibits providing enough of the resource to meet all demands. If two or more resources seem to contend for the honor, pick the resource demanded near the beginning of projects. That leaves you the option to change your mind later if necessary. We can call this the capacity constraint resource because it influences overall company performance. There must be a reason that we cannot easily increase the supply of that resource, which is the company bottleneck and therefore must become the drum for all the projects.

Because the purpose of selecting the drum resource is to stagger the start of the projects and avoid overloading the system, it usually does not matter much if you select the wrong resource as the drum. You will still get some degree of project staggering. As long as you choose a relatively highly loaded resource that you cannot easily elevate, you are likely to get a large benefit. Project performance will help you focus on the correct drum resource over time. It is far better to get on with the drum schedule with the wrong resource than to continue to operate the old way while agonizing over the actual drum resource.

Many criteria have been proposed to identify the drum resource. With project plans, you do have the total resource demand, and you should know your total resource on hand. You could select the drum by the highest ratio of demand to available staff. Use this method only if you have some reason to believe both numbers for all the projects. Dr. Goldratt does not recommend this method for production because he claims the data are never very good. That may also be true for projects. If you use this method, make sure the resource selected is not easily elevated, for example, by hiring contractors or temporary staff.

To achieve the maximum effect of staggering the projects, the drum resource should be the resource that controls the largest amount of critical chain time on your projects. This resource may vary from project to project. If, like many companies, your projects tend to follow a repetitive pattern (e.g., from engineering to construction to operation), you may find one resource that dominates critical chain time. Selecting the drum resource makes it most likely that you will remove resource contention for all the other resources in the project.

Avoid assigning resources by individual name

Many companies choose to identify resources by individual names. They feel that the resources are so highly specialized that they cannot do otherwise. If that is true, you have no other option. I will say that your company is at high risk, however, if your total multiproject throughput is controlled by one or more individuals who, if they leave or get sick, will bring all projects to a halt. Consider this situation as part of your project risk management approach.

The preferred approach is to assign resources by type in your plans and then have the resource manager assign specific individuals as a task comes up to be performed. The definition of a resource type must assure that any person with that designation could do the tasks assigned to that resource type. The primary advantage to assigning resources by type is that the larger the resource pool, the more advantage you have to dynamically assign resources to projects as the activities demand. That applies to all resources, not just the drum resource. You can, when the task allows it, further accelerate tasks by assigning more than one resource of the type to the task.

7.3.3 The drum schedule

The drum schedule is the plan for allocating the drum resource across all projects. It is usually managed by the manager who has responsibility for the drum resource. The drum schedule is the primary determinant of the system capability to process projects. It sets the start date for each project.

The drum manager needs the drum resource demands for each project and the project priority to create the drum schedule. The individual critical chain project plans determine the duration, earliest time, and relative times for each of the drum-using activities in each of the projects. Figure 7.3 illustrates the drum resource demand from three projects, positioned from highest priority on the bottom to the lowest priority at the top. The drum schedule must fit in all three projects while not exceeding the capability of the drum resource, assumed to be two resource units for this example.

Note that the drum resource use cannot be scheduled earlier than shown on Figure 7.3. That is because other activities on the projects have to feed the drum resource using activities. These are the earliest times that the projects could use the drum resource.

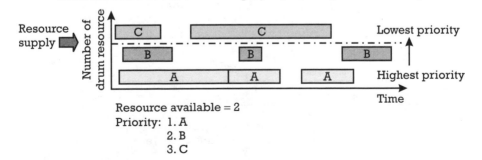

Figure 7.3 Demand of three projects (A, B, and C) for the drum resource, assuming all three projects were to start today. Only two units of the drum resource are available.

The method is to push the lower priority projects later in time until they fall in under the resource supply. That creates the drum schedule. Note that when you are scheduling the drum, the task duration taken from the individual project schedules is the average duration. Because you will want a low risk of not having the drum resource available, you must allow time in the drum schedule for longer than average actual duration. You accomplish that by including the CCB in the drum schedule. Figure 7.4 illustrates the resulting drum schedule.

7.3.4   The capacity constraint buffer

The CCB ensures that the constraint resource is available when it is needed by the project. It is placed between the use of the constraint resource in the prior project and the first use of the resource in the project you are scheduling. It does not take lead time out of the project you are scheduling, but it defines the start date for the resource-using activity.

You size the capacity constraint buffer using the duration of the activity in the prior project. If you have two estimates for that activity duration, the buffer is simply the difference between the two estimates. In other words, the drum schedule allows for the use of low-risk estimates for the drum resource.

7.3.5   The drum buffer

The drum buffer ensures that the drum resource has input to work on when it is needed in the project. In that respect, the drum buffer is a

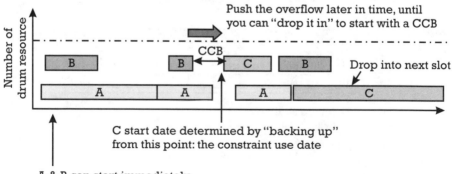

Figure 7.4   Drum schedule accommodates all project demands, including CCBs.

feeding buffer. You place it in the project schedule immediately prior to the activity using the drum resource. It directly affects the project start date and lead time, if it is on the critical chain. The drum buffer is usually on the critical chain, but it is not necessary that it be on the critical chain.

Size the drum buffer as if it were a feeding buffer, using the upstream activity path. You can use the rule-of-thumb sizing method (i.e., 50% of the preceding chain), or you can use the square root of the sum of the squares method (see Section 6.4).

Some have suggested sizing the drum buffer using an arbitrary lead time; such as 14 days. I do not understand the basis for that recommendation, other than it stems from a concern that management may have a tendency to put multitasking pressure on the drum resource. Because a properly sized drum buffer will usually have the activity input ready before the drum resource is available, that may tend to put pressure on the drum resource to multitask or hastily complete the prior task. Either of those behaviors would have negative consequences. Avoid them.

7.3.6 Project schedules

Once you have the drum schedule, you create the individual project schedules by aligning the start of the project to match up the drum-using activity. In other words, you work backward from the drum-using activity to schedule the start of the project. Because you had to have the project critical chain schedule with a time-now start date to create the drum schedule, that amounts to delaying the start of some project by the amount you had to delay the drum resource–using activity to fit it into the drum schedule, plus the drum buffer. You then schedule the rest of the project downstream from the drum-using activities.

7.4 Introducing new projects to the enterprise

New projects can arrive in a multiproject environment at any time. You will have a list of prioritized projects and a drum schedule, and you will know the status of all the ongoing projects. You have to fit the new project into the system.

The only way to schedule a new project is through the drum schedule. To do that, management first must decide where the new project fits into the project priority. It may be the lowest priority, if management prefers the first-in, first-out priority method, or it may fit higher than some of the ongoing projects. For example, if the new project is for an important customer, management may want to place it higher in the priority than in-house projects.

You then must prepare the critical chain schedule for the new project, to determine when (in relative time) it will demand use of the drum resource. You can then fit that resource demand into the proper sequence in the drum schedule. The drum schedule determines the start time for the project by backing up from the time the drum resource will be available for the new project.

If the new project is placed at higher priority than some of the ongoing projects, the schedule of the ongoing projects will change. That can lead to an interruption of work. Use common sense when interrupting project work— for example, do not interrupt nearly completed tasks or tasks that do not have immediate resource demand from another project. Management should consider the potential impact of such interruptions when placing a new project at higher priority than an ongoing project.

Figure 7.5 illustrates the introduction of a higher priority project into a drum schedule. You first put it into the schedule assuming that the project started right away but above the next lower priority project. Put projects of lower priority than the new project above the new project. Then, you fit in the drum use as best you can, as illustrated in Figure 7.6. That may lead to suspending some ongoing projects. If you do suspend ongoing projects, do so wisely— for example, do not stop nearly complete tasks without completing the task result.

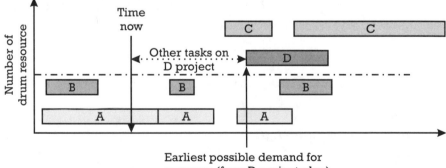

Figure 7.5   A new project (D) is added to the drum demand and judged by management to have higher priority than an ongoing project.

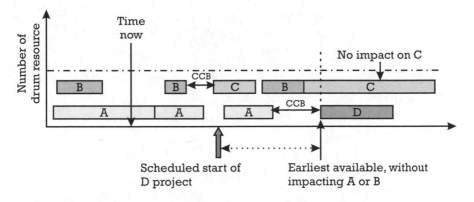

Figure 7.6 Resolving the drum demand sets the schedule for the drum resource in each project, including the new project.

Always keep in mind that the worst possible priority decision is to not make a priority decision, to encourage everyone to "just do your best." That inevitably will lead to multitasking and the worst performance on all the projects.

7.5 Summary

The critical chain for a single project is usually not the constraint for an enterprise performing multiple projects. It is necessary to identify the multiproject constraint, and go through the focusing steps to adapt the CCPM process to firms with multiple projects. When you identify the multiproject constraint and use it to schedule projects, it is the drum for your organization.

- ▶ The drum resource is the constraint in a multiple-project environment.

- ▶ Management must select the drum resource and prioritize all projects for access to it.

- ▶ The CCB ensures that the drum resource is available when it is needed.

- ▶ The drum buffer ensures that the drum resource is not starved for input.

▶ Individual project critical chain plans operate to the start times developed from the drum schedule, including the CCB and the drum buffer.

▶ Management must introduce new projects to the system through the drum schedule by first assigning the priority relative to ongoing projects and then scheduling the drum-using activities.

Practical applications of CCPM have demonstrated the greatest gains in multiproject enterprises. The reason is that those environments usually require everyone to multitask much of the time. Elimination of much of the bad multitasking has the greatest impact on overall enterprise project throughput.

CHAPTER

8

Contents

Measurement and control

Measures drive actions that move you toward the goal. In *The Haystack Syndrome* [1], Dr. Goldratt notes:

> The first thing that must be clearly defined is the overall purpose of the organization—or, as I prefer to call it, the organization's goal. The second thing is measurements. Not just any measurements, but measurements that will enable us to judge the impact of a local decision on the global goal.

For a project in a profit-making company, the project goal has to relate to the company goal, which is to make money now and in the future. In this book, I have presumed that performing a project to meet the customer's needs for the budgeted cost on or before the committed delivery date will support that goal.

For a project in a not-for-profit organization, including both private and government organizations, the general statement of the goal can be converted to performing their mission now and in the future (unless that mission is to eliminate some problem). For that goal to be operational, however, you have to convert it into a measurable quantity so you can dig down and define local measures for decision makers in the organization.

Once a project has begun, the project manager's decisions focus on how to deliver technical quality on time and for or under the estimated cost. Project-level decisions include the following:

- Disposition of material that is not up to specifications (including, for R&D projects, not getting the hoped-for result);
- Requests for additional time or money to complete activities;
- Requests to add scope (someday, maybe even a request to reduce scope!);
- Unanticipated resource conflicts;
- Late activities that may threaten the delivery date;
- Unanticipated external influences like accidents, weather, new regulations, and unfulfilled assumptions (e.g., soil conditions dictating a need to put in pilings before construction);
- Recovery from mistakes.

In the following list of what effective measures must do, the first six were identified by Dr. Joseph Juran [2]. Effective measures must:

- Provide an agreed-on basis for decision making;
- Be understandable;
- Apply broadly;
- Be susceptible of uniform interpretation (i.e., be easy for everyone to understand the same way);
- Be economic to apply;
- Be compatible with existing design of sensors;
- Provide early warning of the need to act;
- Deliver control data to the person who must act;
- Be simple.

The CCPM measurement and control system is designed to satisfy those requirements.

8.1 Buffer management

The measurement system for CCPM follows the practice established in drum-buffer-rope. It uses buffers to measure critical chain plan performance. Section 4.3 recommended explicit action levels for decisions. The project buffer is the most important monitoring tool.

The resource buffer protects the critical chain from resource unavailability. The resource buffer should trigger a two-way communication between the project manager and the resource manager to ensure that the resource will be available when it is needed. The project manager tells the resource manager, "Based on present status, we will need resource x in y days." The resource manager responds, "Based on current resource use, that resource will/will not be available." The resource buffer should be large enough to allow for alternative actions in the event there will be problems with resource availability.

Managing the feeding buffers protects the overall schedule from delays in merging paths, including paths that merge at the project buffer. Do not use resource buffers on noncritical chains. Instead, protect the project from noncritical chain activity delay by both the CCFB and the project buffer. Action criteria for the CCFBs are the same as for the project buffer.

In the multiproject environment, treat the drum buffer like any feeding buffer. You do not have to measure the CCB, which was used to schedule the project start time and not as a dynamic measurement.

8.1.1 Status reporting

Buffer reporting depends on realistic estimates of how many days are left to complete a task. There is often a tendency to report "on schedule" until the due date arrives. With the critical chain measurement system, that amounts to subtracting the total duration estimate from the days spent. You should question estimates that are repeatedly on schedule. A useful aid to estimating is to ask people, particularly on the critical chain or on feeding chains with significant buffer penetration, to explain the basis for their estimated number of days remaining. For example, "This task was scheduled to generate about 300 lines of code in 30 days. The 300 lines still look about right. We generated only 100 lines in the first 15 days, but that included laying out the format and relationships. Therefore, we should be able to complete the remaining 200 lines in well under 20 days; I now estimate 15. We have about 5 days of testing to do at the end. So, we have about 20 days remaining." The more use you make of quantitative measures of task status, and use these to forecast task completion, the more effective your project control.

8.1.2   The buffer report

Clients always want to know how their project is going. Project management usually wants to keep the client separate from the people performing the work for a variety of reasons. Reasons include the clients disturbing the work flow, workers mistaking client input as direction to change the project, and the client receiving inaccurate information by asking people questions to which they do not really know the answer (everybody likes to help).

Most of us are aware of the organization filter effect. I once had a boss tell me he believed that nothing important got through two layers of management. I used to think he was a pessimist. I now think he was an optimist. The same thing happens to written information as it passes up the chain. That is, candid information is often filtered out of written reports. Therefore, clients usually are not content with dealing with formal reports or transmissions through the formal reporting system.

One of the best ways to keep clients directly informed with accurate information is to invite them to your project status meetings. I recommend that every project have weekly and monthly status meetings. The meetings must be highly informative and tightly focused to the needs of the attendees. I recommend a fixed agenda and the rapid publication of minutes following the meetings (hours at most). In addition to the task status, the agenda should include review of actions from the previous meeting, review of the project risk list, and status of any change control requests. Buffer signals should create action items on the action list.

Most projects require some type of formal reporting, most often on a monthly basis. With today's computers and the sophisticated project control programs, it is much too easy to create very large reports. (The cartoon strip "Dilbert" illustrates the problem with large reports by having Dilbert's boss use a thick project report as a footrest.) Project reporting should help the project, not demand otherwise scarce project resources. Therefore, the reports should be focused on the customer's need for the report. Figure 8.1 illustrates a simple format for project reporting. The report should contain the minimum information necessary to meet that need and have a one-page executive summary that tells it all.

The project team is often overlooked as the recipients of project reports. They rarely have the time to read thick reports and often do not

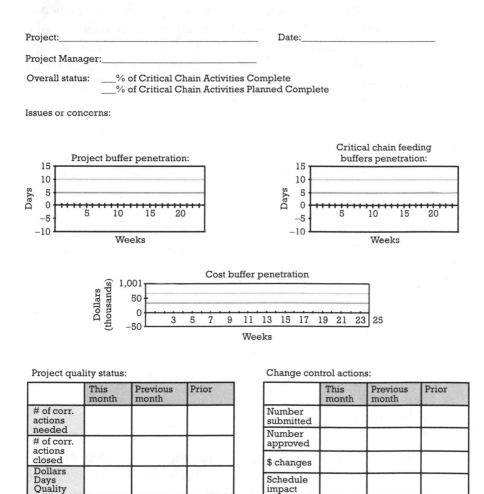

Figure 8.1 Example of a project status report that plots buffer trends.

have access to them. Let's hope that situation is improving with personal computers and intranets; there is little excuse for printing large-volume project reports today. There is no excuse for failing to make the information available to project participants. I recommend that large projects include a formal process of reporting back to the project participants monthly. On a large project, you may not want all of the project team at the monthly project control meeting, but on a small project, it may be appropriate to invite everyone.

Plot trends of buffer utilization are illustrated in Figures 4.9 and 8.1. The buffer measure is functionally similar to a control chart, and you can use similar decision rules. That is, any penetration of the red zone requires action. Four points trending successively in one direction require action. Trending is important if your processes that produce project task results are not in statistical control. Shewhart notes that the trend information is even more important in such cases [3].

8.1.3　Resource use of buffer reports

Resources and resource managers have to see the buffer reports from all the projects they support. Resource managers use the buffer reports to make decisions on assigning resources. In addition to the resource buffer triggers, the resource managers use the buffer reports to decide the priority to dynamically assign specific resources to tasks. The criteria are:

▶ Critical chain tasks have priority over noncritical chain tasks.

▶ For two competing critical chain tasks, the one with the most relative project buffer penetration has priority.

▶ For two competing noncritical chain tasks, the one with the most relative feeding buffer penetration has priority.

▶ For equal buffer penetration, the project with the nearest end date has priority.

Resources can use the buffer reports and the same criteria to make decisions between multiple tasks that have been assigned to them. The overriding rule is to engage in roadrunner behavior for whichever task you work on.

8.2 The cost buffer

For many projects, cost is as important as schedule. For some projects, cost may be an absolute constraint. In such cases, it is useful to extend the buffer idea to manage cost to budget.

Sizing the cost buffer requires considering a number of factors. First, you should account for the fact that you have not budgeted for the use of the schedule buffers. While start delays will not directly translate to cost, additional activity duration times used by people working to complete the activity will increase cost. You should include at least 50% of the time buffers into the cost buffer, at an appropriate cost rate related to the chain they protect. Alternatively, you could add the amounts removed using the sum of the squares method, if you believe that there is no bias in the individual cost estimates. That is, the cost buffer equals the square root of the sum of the squares of the cost removed from each project activity. Note that this method is subject to the same considerations that apply when you use the sum of the squares to size time buffers. For example, for many nearly equal cost activities, this method may yield a much smaller buffer.

Second, you must consider the unique aspects of each project that affect your ability to estimate accurately. For example, if you are estimating unique materials or materials subject to wide price variations, you should consider that when sizing the cost buffer.

Finally, take advantage of using an aggregated cost buffer, which substantially reduces the total cost buffer requirement. It also reduces the tendency to use it or lose it, which sets in if you include cost contingency in each activity. As with schedule, because of human behavior, projects do include cost contingency. The only concern is if you have a readily identified aggregated contingency under the control of the project manager or hidden contingency at the discretion of each task performer.

Never attempt to operate with a cost buffer of less than 10% of the estimated project cost. The reason is that there is always some bias in project cost estimates. You can always forget some things and sometimes underestimate things. Project reviews will usually remove any additional unneeded items in the cost estimate and ensure that individual cost estimates are not unrealistically high.

8.2.1 Cost buffer penetration

Penetration of the cost buffer provides the global information you need to drive cost decisions. The measure is cost buffer penetration in dollars, and the action levels are the same as for the time buffers. Take no action in the first third of the buffer, plan for actions in the second third of the buffer, and take actions when you penetrate the third third of the cost buffer. The cost buffer includes two elements: the net effect of approved project changes and the difference to date between actual cost and planned cost for the work performed.

You cannot compare actual project cost to planned project cost versus time to calculate buffer penetration. The reason is that actual cost to date includes actual schedule performance, and planned cost to date is based on the scheduled activity performance. Because of variation, actual schedule performance never matches scheduled performance. Use the earned value method to determine cost buffer penetration. As described in Chapter 3, earned value was developed precisely to separate out the two contributors to the difference between cost and estimate on a project: schedule performance and cost performance. Earned value defines three terms:

▶ Actual cost of the work performed (ACWP), which is simply how much you have spent to date on the project, broken down to elements of the project.

▶ Budgeted cost of work scheduled (BCWS), which is the time-phased budget for the project.

▶ Budgeted cost of work performed (BCWP), which is the earned value. You credit activities with a portion (from zero to 100%) of the budgeted cost for the activity. (Note that the actual cost to perform the task does not matter to BCWP or earned value.)

The only new term here is BCWP. ACWP is simply the cost to date. BCWS is the budgeted cost to date. The difference between ACWP and BCWS is the spending variance. Spending variance is made up of two parts: the cost variance and the schedule variance. Use the cost variance to determine cost buffer penetration:

$$CV = BCWP - ACWP$$

Most computer scheduling software includes the capability to calculate the earned value, or accumulated BCWP. The ACWP is your actual project cost as of a given date.

Two of the earned value measures are the same as critical path cost measures, that is, the BCWS and the ACWP. Earned value really introduces only one new measure. Most earned value practitioners consider BCWP a schedule status indicator. Unless you have all your budget information loaded into your schedule file, that requires processing the activity status data with the budget file. Normally, that cannot happen more frequently than the accounting system runs, so it usually is monthly. The delay is a problem in many projects, because it turns schedule status into history. It makes the whole job of project management equivalent to driving your car by looking through the rearview mirror.

Most projects have little trouble coming up with comparable BCWP and ACWP for direct labor hours or cost. Many projects and companies do have trouble achieving comparable values for material costs. The problems come from delays, accruals, and commitments.

Few companies are yet able to compile effective actual cost reports more frequently than monthly and more quickly than a week or two after the end of the month. Time lags may be greater for subcontract work. Unless a project is very long, a significant portion of the project time or budget may be expended before the project manager sees it in cost reports. Multiyear government projects have to work to annual budgets as well as overall project budgets; thus, the six-week delay can represent 10% of the annual budget.

Material costs may include contract labor. The reason project control systems have difficulty is that the financial systems often lag actual material expenditures. Make sure you account for that in determining cost buffer penetration.

The accrual problem occurs because you often do not get billed for long lead-time materials as they are built by the supplier; you get the bill upon delivery and usually take a month or more to pay for it. Your schedule system usually spreads the cost for the material over the time from placing the order to delivery, sometimes many months. Your financial system does not account for the cost until it is paid in one lump sum some time after the actual delivery. To account for that, some companies estimate accruals and include them in the project ACWP.

Accruals are estimates of what you owe on the material. Unfortunately, accrual systems are notoriously inaccurate and often have a delay of their own.

Material commitments are the total value of signed contracts not recognized as costs in your accounting system. You may have budgeted $10,000 for some piece of equipment and then signed a contract of $15,000, because that is the best price you could get at the time you placed the order. Your cost variance should include the difference as soon as you sign the contact, because your project will see the cost. In most financial systems, you will not see the difference until the costs are accrued over time, or until the payment is actually made. Some project management systems prevent you from changing the budget to account for the difference. You may have to account for the difference between committed material cost and actual material cost separately and add it to your cost buffer penetration.

8.3 Quality measurement

Ireland describes the fundamentals of project quality management [4]. CCPM does not directly affect the requirements or processes necessary for project quality control. TOC places a premium on process and product quality because of the importance to the company goal. TOC is a process of ongoing improvement. Quality systems such as those prescribed in ISO 9000 are completely compatible with CCPM.

In *The Haystack Syndrome*, Dr. Goldratt described an effective measure of quality as dollar-days [1]. Dollar-days are the accumulation of the dollars of impact for each day an item does not meet the requirements. Dollar-days put the correct focus on quality of product. The technical measures of product quality do not relate directly to either the project or company goal and the necessary conditions. You measure technical performance by conformance to customer requirements, as defined in the customer requirement lists or other specifications, standards, or sources of product requirements. Such diverse measurement units do not give us an understanding of the importance of quality.

The quality assurance function usually has the lead to measure and report on quality conformance to plan, but a preventive approach to quality requires all project participants to plan how not to make defective product. You need a measure that correctly portrays the cost of poor quality and incentivizes prevention of quality defects. Dollar-days encourage quality performance on a project. If one activity must reject an input that does not meet quality standards, the product goes to the quality department, which accumulates the dollar-days until the product is passed on to the activity that caused the defect. (The activity that rejected the product does not get credited with dollar-days.)

Dollar-days can help you provide the necessary incentive to produce quality deliverables from each activity. As soon as an activity completes and passes on the work result, the dollar-days are passed on to the successor activity. The dollar-days continue to grow until the excess activity time recovers. Downstream activities will realize it is not fair to penalize them for the overruns of duration by predecessor activities; nevertheless, in most cases, they will be a little more motivated to get their activity done and pass on the hot potato.

What are the dollars to assign to the dollar-day computation? Several choices come to mind. The daily burn rate of the activity is the low end. The overall project cost is the high end. For activities on the critical chain and for activities on the feeding chains, once they have consumed the project buffer, the total project cost seems appropriate. That will cause immediate focus in the right place. For activities on feeding chains that have not used up the project buffer, the value of the chain would seem appropriate.

8.4 Responses to buffer signals

8.4.1 Schedule buffer exceeds first third

This is a signal that the schedule buffer may have been violated, affecting the overall project schedule. At that level, you must plan ways to recover the schedule on current or downstream activities on the chain. There are four general ways to reduce path time: Change the activity logic, increase resources, reduce scope, or improve the process for the activity. Table 8.1 lists ideas to carry out three of those four methods.

Changing the activity logic usually means putting more activities in parallel. If you choose to do that, be sure to assess the need for additional feeding buffers.

Table 8.1
Ideas to Help Reduce Schedule Buffer Penetration

Methods to Increase Resources	Methods to Reduce Scope	Methods to Improve the Process
Hire additional staff	Subcontract part of the scope	Change activity logic (e.g., go from finish-to-start to finish-to-finish) Examine the activity logic for ways to reduce batch sizes
Break up the activity to use a more diverse kind of staff	Revise requirements	Provide improved tools
Authorize overtime (for labor)	Defer requirements to later in the project	Obtain expert assistance
Subcontract labor	—	Use process improvement tools, especially cycle time analysis

8.4.2   Cost buffer exceeds first third

This is a signal that the overall project may overrun the budget. You must plan ways to reduce cost (Table 8.2). Depending on the trend and the indications and projections from the cost buffer, you may initiate action before exceeding the second third of the cost buffer.

Table 8.2
Ideas to Help Reduce Cost Buffer Penetration

Methods to Increase Resources	Methods to Reduce Scope	Methods to Improve the Process
Use lower cost staff	Subcontract part of the scope	Change activity logic (e.g., go from finish-to-start to finish-to-finish)
Use more productive staff	Revise requirements Negotiate	Provide improved tools
Use competitive bidding for subcontracts	Defer requirements to later in the project	Obtain expert assistance
Perform make-buy analysis on planned subcontracts and on activities that might be subcontracted at reduced cost	Look for activities that can be deleted	Use process improvement tools, especially cycle time analysis
—	Look for costs that may not be necessary to meet the customer's requirements	—

8.4.3 Dollar-days quality increasing

This means that the quality process is not effective. You need to perform problem solving to discover and correct the core problem. You must have the quality process well defined and in control for problem solving to be effective.

8.4.4 Schedule buffer exceeds second third

This is the signal to implement the action you had planned. Depending on the changes necessary to implement recovery, you may need to adjust the project plan, using your formal project change management procedure. If you change the project logic, such as going to finish-to-finish logic instead of start-to-finish logic, change the plan accordingly. Changes such as authorizing overtime or using contract labor should not require a change to your plan.

8.4.5 Cost buffer exceeds second third

This is the signal to implement the action you had planned. You need not change the plan for cases in which actual cost simply exceeded the estimate, and you intend to absorb them in the cost buffer. As for the other signals, if your plan requires changes in the project logic or scope, you should implement a formal project change along with implementing the action.

8.5 The cost world

I discuss cost-world measures because many critical chain project managers will be required to report to a cost-world accounting system, at least for a while. The general point I want to make is that using CCPM will not prevent you from reporting to cost-world measures. Typical measures employed on projects for cost and schedule management include the following:

- Milestone status: Days early or days late;
- Activity status: Projection of ability to deliver on the milestone date, at the budgeted activity cost;

▶ Critical path status: Projection of time to complete the project;

▶ Actual cost versus budget status: Comparison of actual cost to budget cost and estimate to complete the project.

Some project managers use somewhat more sophisticated measures, especially on larger projects. For example, they plot milestone completion on control charts or report and do something with slack time on noncritical paths.

When you are managing to the critical path, it is normal to evaluate the activity status, milestone status, progress on the critical path, and overall cost to overall budget. Usually data are available to monitor cost to budget at lower levels of a WBS than the total project. Activity status is normally gathered in real time by the project manager at status meetings with work package managers. Actual cost data usually have some delay, since most accounting systems run the costs only monthly. The cost data usually come directly from the accounting system, with which managers in the company are familiar.

Recall that when you resource-level a critical path, it is no longer the critical path. The critical path is actually the resource-dependent chains you have created. It is the de facto critical chain. Unfortunately, it usually is not tracked that way after resource leveling. Thus, the critical path project manager loses focus.

Many people apply the critical project management software to update the plan to date. That allows delay in any path to create a new critical path. Worse, it adds the delay to the calculation of float for all the other project paths. Thus, if the project recovers the apparent lost time on the new critical path, it is likely that delay has been introduced in the other paths working to the delayed schedule. Some will argue that comparing the statused schedule to the baseline schedule prevents that undesirable effect. My experience is that most people in such environments focus on the due date listed in the statused project schedule.

CSCSs (cost schedule control systems), introduced in the early 1960s as a method to determine appropriate progress payments on projects, were extended to attempt to resolve a fundamental problem in project management: Cost and time are not in the same units of measure. Although the primary purpose of CSCSs was to provide a basis for contract payments on government contracts, many organizations adopted them as a project management tool. CSCSs define measures for schedule performance (time) in terms of dollars.

CSCSs also compute variances as differences between the numbers and indices as ratios of the variances to the plan measures. I described the cost variance as a useful measure of cost buffer penetration. CSCSs also develop another measure, inappropriately titled the schedule variance, that measures the cost difference due to actual versus planned schedule performance. It is a cost measure, not a schedule measure. While it provides a definition of the other part of the difference between actual cost to date and scheduled budget to date, it has little value as a schedule indicator because it weights schedule performance by the cost of the tasks. It can give entirely misleading information on actual schedule performance. You should not use it.

CSCSs calculate all three of the measures (BCWS, ACWP, BCWP) down to the finest level of detail determined for project measurement and roll them up to the total project. Many of the projects that have suffered the grandest schedule slips and cost overruns have had the most sophisticated CSCSs.

Milestone status (including critical path milestone status) is a direct measure of the project schedule status, available to the project manager immediately as the project is sensed. Most project managers ascertain milestone status at least weekly, either in a meeting or through a simple reporting process. Now that most projects are hooked up on e-mail, the status reports can be compiled in real time.

Putting schedule status in terms of dollars is not meaningful. Schedule variance is further confounded by the ability to account for BCWP a variety of ways and any rollup likely will be a combination of many ways. BCWP makes no distinction between activities on the critical path/chain and those not on the critical path/chain. The actual meaning in terms of the schedule is substantial. As noted earlier, the actual cost impact of activities on the critical chain (the project bottleneck) could be the whole cost of the project for the time lost, not just the cost of the individual activity.

Finally, CSCS usually creates some indices, essentially the percentage of variance. That presumably was to allow constant comparison through a project life and between projects. But there are different ways to calculate that value. One method even goes so far as to recalculate a time from the schedule variance (schedule variance divided by project expenditure rate) to represent some type of average project schedule performance in days. The numbers require more calculation and are meaningless in terms of project decision making.

The primary problem with CSCSs is that they violate the first focusing step. Rather than finding and focusing on the constraint, CSCSs require attention to each activity based on its cost. They are the ultimate cost-world defocusing device.

If you must report to a CSCS measurement system, keep in mind that the measures will show project completion where you show the end of the project budget. Your critical chain project completion is at the end of the project buffer. One way you can reconcile that difference is to put the cost buffer into an activity that represents the project buffer.

With the critical chain measurement system, you monitor progress along the critical chain by the completed critical chain activities and with the project buffer and critical chain feeding buffers. Every day, you gather the data, which have clear meaning, measured in time. You also monitor actual cost as in the critical path method, so data are available as soon as the financial system cranks it out.

8.6 Change control actions

Section 4.4 described the need for formal project change control. The project managers should approve any changes to the plan. You should have a form to aid tracking changes and a numbering system so you can bc sure you are always working to the latest version of the plan. In an ideal world, the plan would never change. In the world most people are coming from, it changes all the time. You likely will be somewhere in between in your initial efforts. You have to decide on criteria that constitute a change to the plan. Following are some thoughts for your consideration:

- ▶ A change in the plan logic (e.g., add a task, delete a task, change to the predecessor or successor on a task) should be considered a change.

- ▶ A significant change in scope of a task (you have to define *significant*) should be considered a change.

- ▶ A significant change in the task resource estimate or in the identification of the resource should be considered a change. It may be necessary to recheck the critical chain.

- Overrun or underrun of a task estimated duration is not a change.

- Overrun or underrun of a task estimated cost is not a change.

- Project, feeding, and cost buffer action triggers may cause plan changes to recover.

Your change control process should operate fast. You may have a change control board, including your customer when appropriate, to expedite change approval.

Keep in mind that you should focus on managing the project to the plan, not on managing the plan. Do not, for example, make changes to your buffers based on actual performance to date.

8.7   Summary

CCPM uses progress along the critical chain and buffer reporting as the primary real-time predictive measurement tool. Consider clients and project team members as customers of your project reporting and control system. Buffer reporting has to be timely to be effective; you should status and report buffers at least weekly and ensure that the information is available to users within a day.

- Weekly project and feeding buffer monitoring and reporting provides a proactive real-time decision tool for project control.

- The resource buffer is a two-way communication device between the project and resource managers to ensure resource availability on the critical chain.

- Resources and resource managers use the buffer report for dynamic resource assignment decisions.

- If cost is important, use the cost buffer to measure and control.

- Buffer management minimizes the negative impacts of excessive project changes using buffer trends and triggers related to statistical process control.

- Conventional project change control methods are necessary to handle scope changes and impacts of special cause variation.

CCPM users have found implementation of buffer management to be relatively simple and very effective.

References

[1]   Goldratt, E. M., *The Haystack Syndrome*, Croton-on Hudson, New York: North River Press, 1990.

[2]   Juran, J. J., *Juran on Planning for Quality*, New York: Free Press, 1988.

[3]   Shewhart, W. A., *Statistical Method from the Viewpoint of Quality Control*, New York: Dover Publications, 1986 (originally published in 1939).

[4]   Ireland, L. R., *Quality Management for Projects and Programs*, Upper Darby, PA: PMI, 1991.

CHAPTER

9

Contents

Implementing the change to critical chain

Many companies that have never introduced change into their organization successfully implement CCPM in a short period of time. These successful companies require less than three months to get all projects planned and synchronized and to begin to see the benefits of improved project performance and reduced stress on project teams. Success stories include all types of projects and a wide range of organization size.

People report significant success implementing CCPM for single projects after having read an earlier version of this book or having attended a two-day introductory training class. Unfortunately, some organizations that claim to have attempted critical chain for one stated reason or another gave it up. That has happened on both single projects and in multiple-project organizations. The following presents a process proven to work in both single-project and multiproject organizations.

9.1 Implementation model

Figure 9.1 illustrates the basic project model to implement CCPM. Implementation is a project. The end vision requires operating to the critical chain paradigm. Implementation plans vary in content and scope depending on the specific organization. For example, a single-project implementation usually involves only the direct project team and usually can be accomplished by the leadership of the project manager. Multiple-project implementation can be much more involved, requiring the active support of all involved project teams and resources.

The first three steps of implementation are the same for all projects: You have to charter the implementation project, gain endorsement of the project stakeholders, and prepare the project work plan. Stakeholder endorsement is the most important part of an implementation project.

People approach process change with caution. The record of successful change is not good. Dalziel and Schoonover noted, "Technocratic leaders … focus exclusively on outcomes without considering the concerns of employees who must implement and sustain change" [1]. Many project managers (myself included) are technocratic leaders and therefore subject to that blind spot. Dalziel and Schoonover go on to note that "this perspective frequently results in short-term gains, unforeseen pitfalls, and long-term resentments."

The technical aspects of CCPM are not challenging to any organization that has basic project management capability. Many organizations

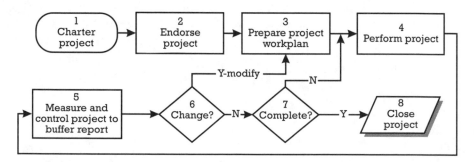

Figure 9.1   Implementation process flowchart.

with rudimentary project management capability have been able to use critical chain implementation as the focus to improve overall project success. CCPM is not so much an advanced project management method as it is a different—and better—method.

CCPM is a management system change. Implementation must address all aspects of the system. Figure 9.2 illustrates the so-called seven-S (strategy, systems, staff, style, skills, structure, and shared vision and values) model (which I understand was developed by McKinzie, but which has been modified by many authors since). Dr. Stephen Covey modifies the model to further emphasize the people part of it and calls it the PS paradigm [2] (the P stands for people). These models provide a system definition for change.

Consider the seven-S model for your organization as you plan implementing critical chain. You mostly must make sure that your change plan does not overlook some facts about your organization that may block implementation or cause unintended consequences. There is no reason to change the structure of your organization to implement CCPM. Organizations ranging from full project to full matrix have successfully implemented CCPM.

The seven-S model is incomplete (deliberately so, as with all models). Deming emphasized the need for managers to consider elements normally considered outside the business system, such as customers, suppliers, and even competitors. Further, the seven-S model is static. All

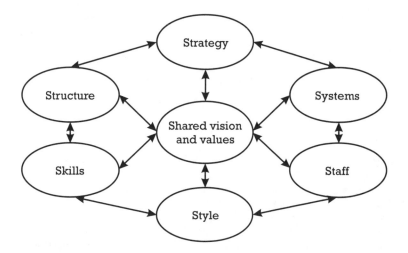

Figure 9.2 The seven-S model provides a system definition for change planning.

business organizations are complex, dynamic, and adaptive systems. That means they are constantly changing. Management's problem is to lead the changes that will happen in a positive direction.

The process to achieve any goal requires effective feedback mechanisms to adjust actual performance to plan. Box 5 in Figure 9.1 shows measurement and control. Buffer management works to determine if you are accomplishing the implementation project relative to schedule, as in any project. Also, as in any project, you need quality results measurements to ensure that the result achieved and passed on from one task to the next satisfies deliverable requirements. In this case, quality results include some soft measures, such as how people are feeling about the change to CCPM and some real dialog on challenges people are having making it work.

Most organizations require behavior changes to implement CCPM. Which changes your organization requires depend on what behaviors your organization currently exhibits. Table 9.1 summarizes typical behaviors and the behaviors demanded by CCPM. Use the tables as criteria to judge how you are doing.

Table 9.1(a)
Senior Management's Behavior Changes

Change	Current Behavior	Future Behavior
Committing only to feasible delivery dates	Sometimes committing to arbitrary delivery dates determined without consideration of system capability to deliver	Committing only to delivery dates with a critical chain plan and (if multiple projects), after sequencing through the drum schedule
Eliminating interruptions	Inserting special requests into the system with no assessment of system capability to respond; sometimes place demands for routine administrative work above project work (e.g., salary reviews)	Prioritizing all requests using buffer report
Setting project priority (only for multiple projects)	Not setting clear project priority or changing project priorities	Setting project priorities, including the priority of new projects relative to ongoing projects
Selecting drum resource (only for multiple projects)	Giving no consideration to system constraint	Selecting the drum resource to be used for sequencing the start of projects and creating the drum schedule

Table 9.1(a) (continued)

Change	Current Behavior	Future Behavior
Selecting drum manager and approving project sequencing (only for multiple projects)	Starting each project independently as funding is available	Drum manager creating drum schedule; senior management approving; project managers scheduling projects to the drum
Project status	Looking over shoulders	Buffer report

Table 9.1(b)
Resource Managers' Behavior Changes

Change	Current Behavior	Future Behavior
Setting resource priority	Assigning resources on a first-come, first-served priority or attempt to meet all needs by multitasking	Assigning resources using the buffer report
Resource planning	Planning resources by name and task	Planning resources by type and assign to tasks as they come up using the buffer report priority
Early completion	Turning in tasks on due date	Turning in tasks as soon as they are complete
Eliminating multitasking	Ensuring resource efficiency by assigning to multiple tasks at the same time	Ensuring resource effectiveness by eliminating bad multitasking
Resource buffers	Resources planned far ahead and not available when needed	Using resource buffers and buffer report to dynamically assign resources to tasks

Table 9.1(c)
Project Managers' Behavior Changes

Change	Current Behavior	Future Behavior
50% task duration estimates	Project managers sending message that they expect due dates to be met	Project managers first getting low-risk task duration and then getting average duration; using task uncertainty to size buffers
Date-driven to task-driven	Providing start and finish dates for each task and monitoring progress to finish dates	Providing start dates only for chains of tasks and completion date only on the project buffer
Feedback on task duration overruns	Management providing negative feedback when tasks overrun due dates	Management providing positive feedback and help if resources perform to roadrunner paradigm

Table 9.1(c) (continued)

Change	Current Behavior	Future Behavior
Project status	Varying; often using earned value as the schedule measure	Buffer report (including a cost buffer)
Project changes	Varying; often submitted to minimize minor variances	When triggered by buffer report
Response to management demands for shorter schedule	Arbitrary task duration cuts	Adding resources or making process changes to get a feasible and immune schedule
Early start	Starting tasks as early as possible	Starting task chains as late as possible, buffered by feeding buffers
Sequence projects (only for multiple projects)	Starting project as soon as funding is available	Scheduling project start using drum schedule
Assigning resources dynamically according to critical chain priority and buffer report	Getting resources as soon as project funding is available and holding resources until they cannot possibly be used any more on the project	Getting resources only when needed and releasing as soon as task is complete

Table 9.1(d)
Subcontractors' Behavior Changes

Change	Current Behavior	Future Behavior
Delivering to lead times	Delivering to due dates	Delivering to lead times
Shortening lead times	Delivering to due dates	Shortening lead times

Table 9.1(e)
Customers' Behavior Changes

Change	Current Behavior	Future Behavior
Eliminating project scope changes	Customers spending little time initially establishing requirements and then introducing late changes	Establishing requirements as part of the project work plan; changing as little as possible with formal change control
Supporting use of project buffer	Customers interpreting contingency as fat	Customers understanding the need for buffers to reduce project lead time and ensuring project success
Eliminating arbitrary date milestones	Demanding arbitrary date milestones	Using plan to set milestones

Box 6 in Figure 9.1 is where you compare progress to plan and determine if you need to take management action. It includes both progress to the work plan and progress to the soft measures. Some companies like to put together a steering group to monitor progress and help the implementation project manager. Box 7 in Figure 9.1 is simply where you mark tasks as complete.

Several features of CCPM make it easier to implement than many changes people attempt to make in organizations:

▶ Nobody loses, because you take away the win/lose aspect of task duration estimates.

▶ Many win, because of the reduction in pressure to multitask and the feeling of accomplishment that follows successful task and project completion.

▶ It is simple, compared with the more-and-more-detailed approach or the implementation of complex earned value reporting systems.

▶ Results feed back rapidly in the weekly project meeting. You do not have to wait for company cost reports.

▶ Unfortunately, not everyone will feel that there is enough of a need to change anything, and others may fear that they have something to lose. Section 9.2 describes some strategies for dealing with such potential obstacles.

9.2 Vision of the end

If you do not know where you are going, you will not know when you get there. Launching critical chain implementation without a clear picture of success is like launching an arrow into the air; it will come to earth "you know not where."

You can represent the end vision in a variety of ways. A picture usually helps; many people respond better to visual stimuli than to other inputs. Consider putting together your own picture of your organization operating to the critical chain paradigm. For engineers, that picture may look a lot like a diagram. To project managers, it may be a picture of a simplified Gantt chart, showing the features of critical chain plans. To people-oriented managers it would include people. I prefer to describe the end vision in terms of the behavior necessary to operate critical chain projects successfully.

9.3   Implementation theory

Change management theories abound. Much of the literature starts with the reasons change attempts fail. Failure means that planned change does not achieve some desired goal. Change goes on in all organizations. It just is not the change wanted by the book authors or the people they asked. The theories rarely agree on the reasons changes fail to achieve the goal. As was the case with project management, that should give us pause to suspect that the authors are missing a deeper systemic cause.

9.3.1   The rule of 3-4-3

The following anecdotal observations are examples of the parables that form the basis for much management theory. A speaker at a recent management conference brought up the rule of 3-4-3. He said he learned it from a Japanese colleague when they were studying the application of *Kaizen* (Japanese for "continuous improvement"), the subject of his talk. The reference means that in attempting to introduce new ideas into any organization, about 3 out of 10 people will catch on immediately and begin to implement. Another 3 out of 10 will remain clueless and uninterested in just about anything and everything, forever. The middle 4 of 10 will behave in exactly the way you might expect middle-of-the-roaders to act—they will wait and see and gradually come aboard, as the change becomes the organizational norm.

The time for organizational change to take place depends, of course, on both the change and the organization. For example, I read that it took the British Army 125 years to accept the recommendation that foot soldiers should wear asymmetric shoes (i.e., one for the left foot and one for the right foot). Just imagine complaints of the supply sergeants: "Hey, you guys want me to be efficient. And now you want me to double my inventory! Think of how hard it is going to be to keep the left and right shoes of the same size together, all the way from manufacturing, through distribution and supply, to the soldiers. Costs will go up. Delivery will be harder to maintain. And the soldiers will never be able to put them on; you have made things twice as complicated!" That argument, or something like it, must have held out for those 125 years.

In science, it is now generally accepted that it takes at least a generation (i.e., 25 years) for a new basic scientific theory to replace the old. (Some suggest at least two generations.) The believers of the old theory have to die before the new theory can take place. Of course, that is for our new and enlightened age. They used to put people like Galileo in jail. Before that, it was "off with his head!" You can understand why the scientific revolution took such a long time to build up steam (in both the physical and figurative meaning of the word.)

Bennis identifies three groups of change strategies [3]:

▶ Empirical and rational strategies, assuming that people will follow their rational self-interest;

▶ Normative re-educative strategies, assuming that change requires alterations in organizational structure, institutional roles, and institutional relationships;

▶ Power strategies, requiring compliance to the leaders' will.

He further describes eight types of change programs that derive from those strategies:

▶ Exposition and propagation;

▶ Elite corps;

▶ Human relations training;

▶ Staff;

▶ Scholarly consultation;

▶ Circulation of ideas;

▶ Developmental research;

▶ Action research.

All eight types of change programs as well as the three strategies seek to use knowledge to gain some desirable end. Most of the strategies rely on rationality. Bennis notes that knowledge about something does not automatically lead to effective action. He concludes by observing:

Sometimes, the changes brought about simply fade out, because there are no carefully worked out procedures to ensure coordination with other interacting parts of the system. In other cases, the changes have backfired and have to be terminated because of their conflict with interactive units. In any case, a good deal more has to be learned about the interlocking and stabilizing changes so that the total system is affected.

In recent years, business change advocates have often focused on the need for organizational culture change to precede significant performance improvement. They often suggest that it takes many years, perhaps 8 to 12, to accomplish significant culture change in organizations. That is one of the reasons a lot of businesses fail before they can accomplish necessary change.

None of those descriptions, while possibly completely correct, reaches even the correlation level of scientific thinking. For every anecdote someone relates, someone else can provide an alternative story that is the exception that proves the rule. Many companies grow to very large size within the shortest time frame discussed (8 to 10 years).

9.3.2 Appreciation for a system

Considering an organization as a dynamic system moves our thinking beyond correlation and into the realm of scientific thinking. You can use a model of the current system to determine what changes will affect the system the way you want. Dynamic models are important because business systems are dynamic. The laws of the fifth disciple (discussed in Chapter 2) apply. One of the most important and difficult to appreciate laws is that causes and effects are displaced in time and space. That means the effect you observe in Milwaukee today may be due to some management action taken in Tampa last year, not due to the new manager that just came aboard in Milwaukee. The new manager simply correlates in time and space with the effect you are observing. No one seriously believes that the outcome of the Super Bowl causes the stock market to do anything, but every year the media discuss remarkable correlations.

Correlation of effects in dynamic systems makes determination of cause difficult. The definition of cause and effect is that the effect invariably follows when the cause is present. (The effect may also be present without the cause in question if it can also follow from additional causes.) The cause of effects in dynamic systems most often is the system structure, not a specific event. Most people have difficulty gaining an intuitive appreciation for that.

Consider chickens and eggs. The question "Which came first?" is meaningless in a dynamic system that includes chickens and eggs. They coexist. Their numbers correlate in time. That is, everything else being equal, the more chickens you have, the more eggs you get. It is true that chickens cause eggs. It is true that eggs cause chickens. Thus, entity causality is completely circular, which is fine in a dynamic system. Based on the "Which came first?" cliché, it does not appear to be intuitive. Depending on the system, it may or may not follow that the more eggs you get, the more chickens you get. Someone may be eating a lot of eggs. That would be part of the system structure and would significantly affect the number of chickens over time.

The thinking process that Goldratt recommends models reality with a cause-and-effect tree, the CRT. This tree structure allows for the inclusion of nonlinear effects and dynamic feedback. It does not provide a way to test and understand the relative importance of system entities and relationships dynamically, but treatment of the feedback loops ("more and more" entity statements in the model) attempts to provide a qualitative understanding of the impact. The method helps plan a move from current reality to a desired model of future reality (the FRT) by identifying a core conflict. The core conflict plays a central role in many of the current system's undesirable effects. The conflict is one of the driving forces that maintain the system in equilibrium. Goldratt's thinking process always provides a starting point for planning change and usually selects one of the more influential parts of the system to begin the change. Some system thinkers identify the influential parts of the system as leverage points. They usually involve a feedback loop. The most effective feedback loops in organizations involve the performance measurement and reward systems.

The core conflict often involves a measurement or policy of the system. It always influences the behavior of people in the system. The thinking process method then surfaces underlying assumptions to identify a starting point to modify the system. Eventually, the process looks to install feedback loops into future reality to accelerate movement to future reality, and to maintain the system in the new equilibrium.

General system theory and system dynamics teach us that feedback loops are one of the most important elements for understanding and influencing system behavior over time. Feedback loops are the forces that maintain the system in equilibrium and can be used to drive the system to new equilibrium. Measurement systems make up the primary feedback loops that drive business system behavior.

9.3.3 Resistance to change

Resistance to change is an essential feature of any stable system. (Please read that sentence three more times.) Open systems are only temporarily stable because the dynamic forces acting on the system, both internal and external, are nearly in balance.

Resistance to change is not inherently good or bad. You can judge resistance to change as good if you want to maintain certain characteristics of a system. For example, you may be pleased that your system maintains a focus on customer service through good times and bad. You may judge resistance to change as bad if you are attempting to eliminate undesirable behavior or to move to new levels of performance. Regardless of your judgment on the matter, the system will naturally resist change.

Figure 9.3 illustrates just a few of the interrelated forces that exist in any business system. Forces are both internal and external to the business system. The forces themselves are interrelated in a complex system structure. Attempts to change any part of the system affect all parts of the system to varying degrees. Because of the linked structure, the net result of those forces will tend to restore the system to its previous state following any disturbance.

Organizational resistance to change is often difficult to distinguish from individual resistance to change. When things are not going as hoped or not going fast enough, people often want to search out and motivate the guilty. Unfortunately, such searches are fruitless. How many people do you know who really want to lose weight, quit smoking, or change some other personal behavior, but seem unable to do it? Or, if they were able to do it, were unable to sustain the progress they made? Do you really doubt their desire or motivation to make the change? Do they not have the skills? Will haranguing them more cause it to happen? Sending them to training? Fredrick Hertzberg observed, "When a man kicks a dog, it is the man that is motivated."

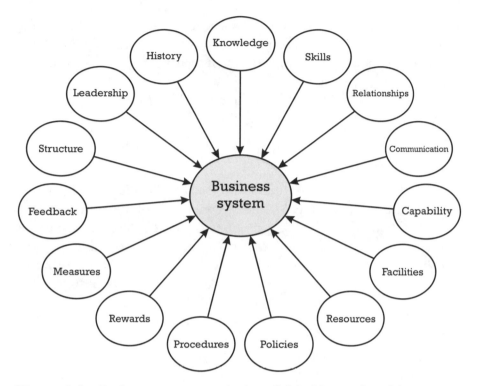

Figure 9.3 Business systems exist in a field of interrelated forces, which naturally push back on attempts to change the system.

The obstacles to organizations making change are the very things that makes them what they are in the first place. The structure of the system determines the reaction that will happen when you try to push a stable system in one direction. You will activate the restraining forces that helped keep the system in balance where it was.

For example, consider an organization that wants to become more efficient. It may choose to eliminate excess resources. TOC teaches that an efficient system can maintain only the constraint at full efficiency. All other resources must have protective capacity to operate the system efficiently. In other words, all other resources must operate at lower efficiency, so the system can operate at maximum efficiency. Unless the company has a good grounding in TOC, it will not understand the necessary protective capacity and will cut into necessary capacity. That will make its system less efficient. The system will resist the improperly imposed attempt to change it. In some cases, due to some of the laws of system dynamics, the system may appear to be more efficient for a few quarters. That is because there was excess inventory in the system, which can make up for the haphazard cutting of capacity. Once that is used up, the system will begin to fail.

You have to consider resistance to change at both the individual and the organization levels.

9.3.4 Psychology

Several properties of the human mind lead to individual behavior that seems to resist change. B. F. Skinner describes one of the more powerful mechanisms. Skinner asserted (with extensive scientific data) that much human behavior comes from what he called operant conditioning [4]. Put simply, that means you continue to do what gives positive rewards, learn to avoid doing things that do not lead to positive rewards (reinforcement), or do things that help you avoid negative rewards. Skinner notes, "A reinforcing connection need not be obvious to the individual reinforced."

Figure 9.4 is my rendition of a control system view of Skinner's model. It starts with a need, which is influenced by the person's current state, including deprivation or satiation relative to the goal. Comparison of that need to the person's understanding of his or her current situation (perceived reality) yields a gap that, if large enough, motivates a person to action. Action seeks to change reality to close the gap. The sensor, which may be the five senses or more removed methods of gaining data, feeds back information about the effect that the action has on reality. If the change is positive (reducing the gap or otherwise supplying a reward), it strengthens the chances that the person will repeat the behavior.

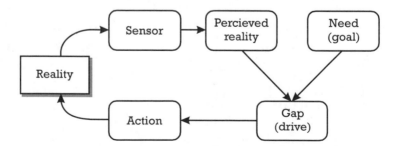

Figure 9.4   Control system view of human actions (behavior).

Operant conditioning must somehow be stored in the brain. Because it defines a (perhaps rudimentary) model of the world (if I do this, then I get that), you can consider it a belief about how the world works. Such beliefs may be conscious or unconscious. Research demonstrates that these beliefs have other impacts on the model. Figure 9.5 illustrates that beliefs affect what you pay attention to, how you interpret what you sense (perception), what your motivations (needs) are, and the decisions you make on how to act in the world so as to increase rewards and decrease negative reinforcers. That influence is mostly unconscious. In other words, you see it because you believe it.

The modern view is that our minds operate as pattern recognition devices. We have a wonderful ability to infer the automobile in the picture by looking at only a small fragment of the picture. We often can name that tune in three notes. It is remarkable, when you think about it.

Section 2.2.3.2 described how beliefs act to focus our attention, and adjust our perception of reality by acting as a kind of information filter. It is reasonable to assume that, at any point in time, people operating in an environment have tuned their behavior to the environment. Put another way, feedback through operant conditioning causes them to behave in

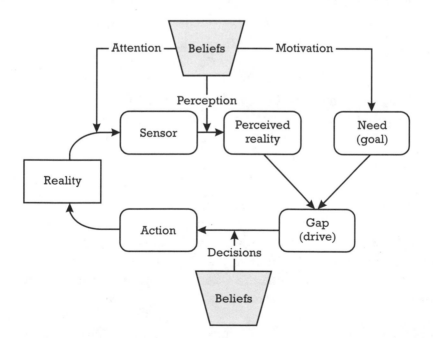

Figure 9.5 Beliefs affect many parts of our internal control process.

ways that maximize rewards in the current environment. Changes in the reward system threaten that position. Furthermore, Skinner demonstrated that extinguishing behavior established by operant conditioning can take a long time. The organism will continue to emit the old behavior, which is no longer reinforced, sometimes for thousands of tries.

Other aspects of psychology, or how our minds work, are also important to understand the system you are attempting to change. One is the availability bias. Psychological experiments repeatedly demonstrate that people are relatively poor judges of probability. They focus on the information they heard or saw last or what impressed them the most, instead of the probability itself. For example, you may hear statements like, "All scientists (programmers, engineers, etc.) tend to underestimate how long it will take to do a task." When pressed for data, people admit to having little. Data analyses often prove otherwise. Most project tasks are reported as complete on the due date (a miraculous occurrence, by the way, proving the existence of date-driven behavior). People also tend to be very overconfident in their ability to estimate probabilities.

9.3.5 Paradigm lock

A paradigm is a belief. People often hold beliefs at such a fundamental level that they do not even realize they hold the belief. Hidden paradigms, or beliefs, influence our thinking. Consider a simple one that many people pay little attention to. The normal practice in critical path project planning is to create the critical path schedule and then perform resource leveling. Most people accept that practice without question. They believe it to be acceptable, because they were taught it as the method to plan projects. It seems logical and harmless. They do not think to question it. After all, if you spend the time to question everything, you will not have time to learn anything. However, they should have asked these questions:

▶ Why does a critical path plan connect tasks in a project schedule only by output to input to start with? The tasks need both the predecessor input and the resource to begin the task. They are equally important.

▶ What happens to the critical path after a project plan has been resource-leveled? Is it the same tasks as before? The previous critical path through these tasks now has slack, because time has been added to other paths by the resource being given to the critical path task first. The definition of the critical path was the path with no slack.

Many people recognize the second question. People find that resource leveling significantly extends the project schedule, sometimes beyond the due date. Therefore, they choose to go back to the nonleveled schedule, with the thought that they can somehow make it work. They then identify one of the causes of project failure as "insufficient resources."

A paradigm shift is a change in a belief. Paradigm shifts, when they happen, happen suddenly. People shift a belief from one pattern to another in an instant. Unfortunately, paradigm shifts do not happen with ease. People are often stuck in a paradigm lock. They can understand the benefits of changing their behavior—perhaps even their belief—but somehow they just cannot do it. Project managers and resources know that the durations people put into schedules are highly uncertain. It is illogical to assign start and stop dates to hundreds or thousands of tasks linked together in a schedule. Nevertheless, they just cannot buy the idea that you can live without doing that. They are locked into the deterministic date paradigm.

Ted Hutchin, an associate of the Goldratt Institute who resides in the United Kingdom, studied the issue of paradigm lock in business as part of his master's thesis. He built the evaporating cloud to describe the dilemma faced by many managers who feel a need to implement change. He interviewed over 350 people in 14 companies, involving 40 to 50 individual clouds, to develop his understanding.

Figure 9.6 illustrates Hutchin's paradigm lock cloud. Reading from A to B, "To achieve my goals, I must maintain control of my life. To maintain control of my life, I must maintain my current paradigm with respect to X." Reading the lower branch, "To achieve my goal, I must overcome the constraints that block me. To overcome the constraints that block me, I must change my paradigm with respect to X." Xs to fill in the cloud include the following:

1. People (scientists, engineers, programmers, etc.) always underestimate how a long a task should take.

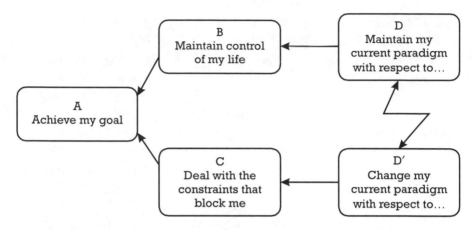

Figure 9.6 Hutchin's paradigm lock cloud describes the dilemma many people face.

2. I am likely to be punished in my performance review if I deliver after the scheduled due date of my task.

3. There is no credit for turning work in early.

4. The sooner the team starts work on my project, the sooner they will complete my project.

5. More detail improves the accuracy of my project plan.

6. Projects need to identify resources by name in order to have people accountable.

7. Project plans have to eliminate resource contention across all projects to ensure that the projects can complete as planned.

8. There is no way to plan for additional demands, such as maintenance of prior product releases.

An important feature of the paradigm lock cloud is that the cross connections, that is, B-D' and C-D, conflict. I feel that I cannot maintain control of my life if I change my current paradigm. I generally can intellectually accept that I cannot deal with the constraint the blocks me if I maintain by current paradigm. Thus, I am stuck with nowhere to go.

The paradigm lock is one reason that implementation of critical chain may fail to succeed. People may gain the knowledge of critical chain, have the necessary skills to perform the behavior, and yet not take action.

The primary fear that managers express in many environments is that they will not be able to get the team or other managers in the organization to make the change with them. They say, "We don't have the time to implement critical chain and meet *our* manager's demands!" They fear they will be out ahead of everyone else and therefore subject to failure, embarrassment, and punishment. It is certainly correct that the first attempts at implementing the changes demanded by the CCPM will be somewhat unstable at first and not as successful as later implementation, when people have had a chance to experience what correct behavior really is.

The primary fear that resources express in many environments is that management will not really change and will criticize or punish them if they are late on project tasks. They also fear that management will attempt to exploit (in a negative sense) any improvement in worker effectiveness, putting them right back where they were before, perhaps with a larger workload.

There is no magic way to break the paradigm lock cloud. Psychological research reveals some ideas that work some of the time. For example, most authors agree that technical discussions have little effect. Many agree that metaphors work more frequently. A body of evidence supports the effectiveness of metaphors. The biggest paradigm change in all of human history, the growth of Christianity, was supported by the use of parables, which are metaphors. The same may be true for the growth of other large groups, because of brain physiology and psychology. The reason metaphors seem to work is that they sneak by our internal perceptual features that strive to keep our internal belief system intact. Dr. Goldratt once used the metaphor of a parachute jumper at the door of the airplane. The jumper had gone through all the training, including tower jumps on a line. But it is a whole different matter to jump out the door of the airplane at a high altitude.

Class exercises can help with weakly held paradigms. I lead an experiment in classes to demonstrate the ineffectiveness of multitasking. *Everybody* gets it. However, that does not mean it always achieves the desired behavior change. Exercises do not overcome the fear that attends making the change in the real work environment. Exercises do not overcome the daily feedback of the work environment. The one thing that consistently works to implement critical chain is for the leaders to lead. That does not mean jump out the door and hope the team follows. (They will not!) It means to join hands and jump together.

Culbert addresses paradigm lock as a major cause of apparent resistance to change, noting: "What is fundamentally necessary is that you understand that distinct interests and motives exist and are the driving force behind people's participation and that these are neither known to you nor under your control" [5]. In other words, you may not have a general case of paradigm lock, you may have many cases with different paradigms. Culbert describes the hypothesis underlying his thesis as the artifact of mind insight: "Organization is an artifact of the mind that views it." That hypothesis is in line with Figure 9.4. As a direct consequence, he suggests, "the people who are targets of the advice are less inclined to experience advice as valuable counsel and more inclined to see it as resistance and self-interested and agenda-biased opposition to how they want to proceed." In other words, it is you who is resisting, not them! Culbert concludes, "There is no formula for producing this type of change, no matter how vivid and compelling the data and life situations you've got at hand." In other words, there is no repeatable solution to paradigm lock.

Dr. Goldratt's evaporating cloud is a good tool to unearth the underlying beliefs or mindsets that separate agendas. It is not infallible, and it does not work all the time. But it works often and is easy to use once you understand it. It is one way to break serious paradigm lock that threatens your implementation.

Figure 9.7 presents a general version of the evaporating cloud in terms of beliefs and actions. (I have come to understand this is as the most basic representation of the evaporating cloud.) The cloud describes two views of reality or two arguments (in the logical argument sense). Consider D and D' as two conflicting propositions about how to achieve the goal. One argument is, "To have A, I must have B. To have B, I must have D." The other argument is, "To have A, I must have C. To have C, I must have D'." Thus, even with a common goal, there are two logical ways to get there. The beliefs may be compatible with each other, or they may not. The actions are not compatible. If they were, there would not be a conflict.

The process to resolve the evaporating cloud is as important as the construct. Usually one side constructs the cloud, with a pretty clear view

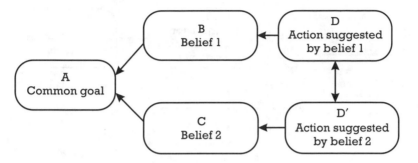

Figure 9.7 Goldratt's evaporating cloud provides a tool to expose and resolve different paradigms.

of what the alternative actions are (e.g., D, D'). The constructor can usually come up with a belief that connects the proposition. They can only guess at the other side's belief. (As noted, neither side may really understand its underlying belief.) The constructor presents the cloud to the other side, reading the other side first. When reading it, they make it clear that they only guessed at C and accept any revision proposed by the other side. The constructor then reads their side, noting, "No wonder we have a disagreement," and then suggests, "Let's search for solutions that will give us A, B, and C and not worry about D and D'." That is a win-win solution. Let's try to identify some assumptions that underlie the arrows in Figure 9.7 and see if we can come up with a way to invalidate one or more of those assumptions and get to our win-win solution."

From here, a concrete example may serve well. CCPM theory asserts that project plans should have relatively few tasks, certainly no more than a few hundred. A common assertion is, "We need more detail in our project plans." Figure 9.8 illustrates one evaporating cloud to specify that conflict.

Read the top branch of the cloud, "To have successful projects, we must have plans consistent with our estimating uncertainty. To have plans consistent with our estimating uncertainty, we must have less detailed plans." Read the lower branch of the cloud, "To have successful projects, we must have control over all parts of the project that may affect success. To have control over all parts of the project that may affect success, we need to have more detailed plans."

Now, begin the search for assumptions that underlie the arrows. Assumptions exist under all the arrows. You read assumptions this way:

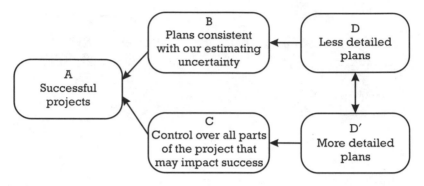

Figure 9.8 The evaporating cloud representing a conflict on the level of detail required in a plan.

"To have A, we must have B *because* of *assumption*." You are limited only by your creativity in coming up with what the assumptions are. A list of examples (only one per arrow) follows.

- ▶ B-to-A arrow: Plans with more detail than our accuracy diffuse focus on the important items.

- ▶ D-to-B arrow: Our estimating uncertainty is on the order of tens of percentage points, indicating plans should only have tens of tasks, and our plans already have hundreds of tasks.

- ▶ D-to-D' arrow: We cannot have both more detailed plans and less detailed plans.

- ▶ C-to-A arrow: Missing details are what makes projects fail. (The devil is in the details.)

- ▶ D'-to-C arrow: More detailed plans improve control.

Once you have the list of assumptions, you can search for solutions that would make the assumption false.

Sometimes, you find that the assumption *is* false and needs no action. If both sides agree, the cloud evaporates. Other times, there is value in defining the opposite of the assumption as a "flying-pig" injection. Flying-pig injections are solutions that would work, but you do not know how you would cause them to happen. (It is like trying to teach a pig to fly. It does not work, and it annoys the pig.) You have to "ground" the flying pig. An example might be, in response to the third assumption

(D-to-D' arrow): "We have both more detailed plans and less detailed plans." It may not be obvious how you might accomplish that. One way would be to have hierarchical project plans, where the plans at any level do not contain excessive detail relative to the uncertainty. This "grounds" your flying pig.

The presenter of the cloud does not suggest the injection to break it. The purpose is buy into a win-win solution. That is best achieved when the target person comes up with a solution you find acceptable, that is, it satisfies the common goal and your belief. If they come up with it and you like it, nothing more is required to make it so. This method often works to break paradigm locks.

9.4 Goldratt's resistance model

Dr. Goldratt developed a model he calls the six layers of resistance to describe the personal aspects of resistance to change. It is a powerful model when considered in the context of the thinking process, or system analysis, that precedes how to cause the change. The model supplements—not replaces—the system analysis. Section 2.3.5 listed Dr. Goldratt's six layers of resistance. The 3-4-3 model generally describes how intensely people will experience the layers of resistance.

9.4.1 Overcoming layers 1, 2, and 3

A large history of learners demonstrates that a two-day training session is sufficient to bring most people through layer 3. Class exercises that enable people to experience, in a laboratory setting, the impact of the critical chain paradigm seem to be an essential part of the success of the sessions.

A small number of people seem to struggle with layers 4 and 5 during the experiential sessions, but they do not seem to revert to layer 1. They remain stuck at layer 4 or 5.

9.4.2 Overcoming layer 4

Layer 4 always comes out just as soon as people begin to create their first real critical chain plans for their organizations. Despite counsel in training to act otherwise, many people go out and ask for 50% estimates from the performing resources. They then find that their critical chain plan is longer than their (resource-unleveled) critical path plan and beyond management's due date *before* they add the project buffer. That leads to consternation and often stops progress.

The project planners put themselves into the pickle. I caution them in training that people do not know the real difference between a 50-50 estimate and a low-risk estimate, which experience has demonstrated many times. For that reason, people will give you an estimate when asked and characterize it as the type of estimate you asked for. If you start by asking for 50-50 estimates, people will believe that is what they gave you. You will now have a difficult time getting their agreement to a plan with shorter duration than the ones they have framed in their minds as 50-50 estimates. This is a case where you really do go all the way back to layer 1 of resistance.

People often bring up fears about potential unintended consequences of implementing critical chain as a reason not to proceed. I encourage them to consider such potentials and to then design a way that prevents or mitigates the unintended consequence. However, experience demonstrates that it is not a good idea to plan to prevent all potential occurrences of unintended consequence before you start. As President Eisenhower said, "Nothing would ever be attempted if all objections had to first be overcome."

For example, a frequent concern is that if you succeed to implement multiproject critical chain, some customers will receive more of the benefit of project acceleration than others will. Even though all the projects will get done quicker than any did before, some people are concerned that less acceleration will cause some customers to be unhappy. I originally marveled at expressing that concern, because the only implied solution is to not implement critical chain, that is, to ensure that all customers get their results equally late. Of course, there are other solutions.

You might now ask, "Are there ways I can make use of my newfound ability to deliver early and satisfy all my customers?" Most people, after getting over the shock of having to respond positively to a negative concern they raised, are able to think up several ways. For example, offering acceleration at a premium price or simply beginning to deliver early against normal lead times. Such strategies amount to market segmentation.

Some customers judge responsiveness by when you start projects rather than by when you finish. Contractors to the U.S. government most often express those concerns. In such cases, the client gets involved in many details of project performance. Many contracts, for example, require data submissions *x* days after the contract has been signed. Once again, having surfaced, this concern gives you the information you need to prevent or mitigate it from being a problem for you or your customer. It is never a reason to forego the benefits offered by CCPM.

Although I always deal with the issue in two-day training sessions, layer 4 usually comes out again during implementation as some form of "What if (management, the client) cuts my buffer, takes away my buffer, or simply judges my success relative to the start of the buffer?" Successful implementations have never had that problem in reality, but explaining that to people standing on the precipice does not bring sufficient peace of mind to unlock their paradigm and cause them to leap.

9.4.3 Overcoming layer 5

The most common manifestation of layer 5 is that management will not change. I dealt with this with a client as I was writing this paragraph. Management has a system of quality gates, or milestones, that must be achieved to complete their projects. They also have a reporting system to inform management of progress to those milestones. The reporting system requires specific dates for each milestone and reporting to each date once a month. The response to my suggestion to take this to management and point out the inconsistency with the critical chain paradigm was met with, "I am not going to suggest to them that they change their measurements. They won't listen to me." That person is now on a path to overcome that obstacle, successfully I am sure.

Equally often, it is the customer who will not change. Or it may be the regulators. Often, it is other organizational elements. The evaporating cloud is a useful tool to overcome the obstacles that become real during implementation. Consider going back to President Eisenhower's view—nothing would ever be accomplished if all objections had to be overcome first.

9.4.4 Overcoming layer 6

Some have suggested that the first five layers of resistance are the active phase, in which people feel safe to raise objections. After all of the objections have been answered, they have no recourse but to go into passive resistance. That is, they seem to support the idea, they just do not do it. People use what Argyris called "skilled incompetence" to avoid change (which he called "organizational learning") [6]. For example, they make certain things are implicitly undiscussable in the organization and then cover it up by making its undiscussability undiscussable. Argyris reports that his research shows nearly all people hold these theories-in-use and that:

> Individuals may unknowingly provide us with distorted information, and those same individuals may hesitate to engage in the dialogue that is required to explore the possibility of such distortions. If we persist in exploring these issues, practitioners may become defensive—this defensiveness leading, in turn, to new distortions, both recognized and unrecognized. [6]

The most effective means of combating layer-6 resistances is to just move ahead and listen a lot. When leadership follows the implementation plan and expects others to follow it, following the CCPM behavior patterns has not been difficult. You have to start the positive feedback loop of project success. The feedback loop will sustain the projects, and that will sustain the implementation.

9.5   To pilot or not to pilot?

The second most common response offered by organizations considering changing to critical chain is, "Let's try a pilot project." (The most common is, "Can you show evidence of companies just like us that have tried this and succeeded?" That from people who want to become leaders of their industry, just as long as they follow everyone else. Go figure.) They hope that the pilot project will help them reduce the risk of full-scale implementation.

Figure 9.9 illustrates the pilot project evaporating cloud. The upper branch supports the desire for the pilot project, while the lower branch supports not doing the pilot project. Some of the assumptions under the arrows of this cloud include:

  ▶ B-to-A arrow: There may be something unique about our environment that invalidates the theory and the experience of others.

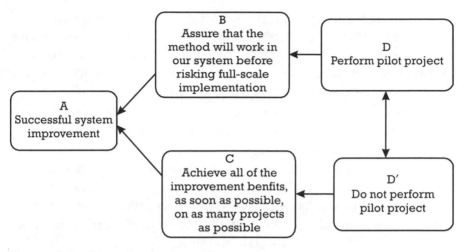

Figure 9.9 The pilot project dilemma.

> ▶ D-to-B arrow: We cannot anticipate or work our way through the things that might block our implementation or cause unintended consequences.

> ▶ D'-to-C arrow: The pilot project does not allow for the multiproject benefits.

> ▶ D'-to-C arrow: The pilot project will delay the benefits to other projects.

> ▶ D'-to-C arrow: The pilot project will not cause behavior change for resources supporting both the pilot project (with CCPM) and other projects using past practice.

Pilot projects sometimes succeed. They can succeed for several reasons, including that the immediately preceding projects were such dismal failures that sheer chance (or regression to the mean, for the statisticians among you) makes it likely that the next project will succeed, no matter what you change or do not change. In addition, you may witness the Hawthorne effect, named after the Western Electric Company's Hawthorne Works, near Chicago, where some early scientific management experiments took place. It describes the response of people broken out and given special treatment. The experiments changed things about production. Whatever they changed seemed to improve performance.

Even undoing the things they did previously seemed to cause increased productivity. Industrial psychologists attribute that response to the desire of a team that is signaled out for special treatment to try harder to give management what they want. In other words, if the pilot project succeeds, people are justified in saying, "So what?"

When pilot projects do succeed, it is still necessary to implement CCPM for the other projects in the organization. In a multiproject organization, the largest behavior changes have yet to be encountered.

You can use the preceding cloud and assumptions, or you can build your own cloud using your interpretation of B. In either case, use the evaporating cloud method with the pilot project proposer. That is, read them the upper part of the cloud first and ask for their words to revise B. Then read the lower part and state the dilemma clearly. Then ask for help on a logical resolution, showing how to develop and read assumptions. Finally, solicit suggestions (alternatives, or, in TOC language, potential injections) that may fulfill both B and C. Go prepared with injections that you would find satisfactory, so you can accept them if the other side suggests them. Do not, however, offer your injection. If the other side does not come up with an idea right away, leave it with them for a few days before you come back to it. Then solicit their thoughts again.

Several potential injections may lead to win-win solutions.

▶ Use the critical chain implementation project itself as the pilot project.

▶ Perform the pilot project in the spirit of plan-do-check-act, that is, a rapid prototype as part of the plan for full-scale implementation.

▶ Invest buy-in effort at the beginning to identify potential concerns unique to your environment and plan preventive or mitigating actions as part of the implementation project risk management plan.

9.6 Plan the change

9.6.1 Endorse the implementation project

Endorsement means getting the stakeholders to agree in the beginning that they are willing to assist as necessary to effect the change to CCPM.

While in some instances it may be enough to have people say, "I don't have a problem with that," in most cases you need more than permission. You need a willingness to change on their part. Stakeholders include the project teams, project managers, resource managers, senior management, clients, and suppliers. There may be more stakeholders important to your implementation; perhaps even stockholders. Sometimes, you may want to obtain the endorsement before you have the project charter. In other cases, you may want to use a draft of the project charter as your vehicle for endorsement.

9.6.2 Charter the implementation project

Figure 9.10 is a sample critical chain implementation project charter. Try to keep yours to one page and focus it on the needs of the project stakeholders.

9.6.3 Create the implementation project work plan

The project work plan is the next step after the project charter and includes the following:

- Detailed specification of the project scope;
- A WBS to organize the project scope;
- Assignment of responsibility to the WBS;
- A resource-loaded (critical chain) project schedule;
- The project budget;
- Definition of the project team;
- Procedures for operation of the project team;
- Plans for project closeout.

Figure 9.11 illustrates the WBS for the project to implement CCPM. The WBS reflects the changes necessary for CCPM. The PRT considered both resource behavior and the technical injections of critical chain, including the following:

- Project plans follow the TOC paradigm (i.e., 50% task times, critical chain, and properly sized buffers).

Project Charter
Project: Implement Critical Chain Project Management
Revision: 0   Date: 2/6/99
Approved by:_____

Project Purpose
The critical chain project management (CCPM) implementation project will install CCPM for management of all projects performed by the Southwestern Division of ACME Products Supply Corporation.

Customer and Stakeholders
The primary individual customer for this project is Wiley E. Coyote, director of ACME Products, Southwestern Division. The customer group is all employees, including managers, of the division. Client customer involvement, such as R. Runner, can be included in this project if client involvement is necessary to implementation.

Project Team
Cynthia Standish is the project director. She will select three to five team members, as necessary, to assist in planning, scheduling, and other implementation project activities. All managers in the division are to support the implementation project as required.

Scope
This project includes all the planning, procedure development, training, and software tools necessary and sufficient to install CCPM into the division. It does not include technical work on the projects nor work with the project customers.

Schedule
The use of CCPM is expected to be substantially complete within 90 days of the approval of this charter. Quarterly progress reviews are to be held for the following three quarters (i.e., the final one on February 6, 2000).

Cost
The overall cost of this project, including expenditures for training (not including employee time), consulting support, procedure development, and the software tool, shall not exceed $250,000 without additional management authorization. Cost associated with the replanning of projects using CCPM and buffer management are not included in this cost, because they are part of the respective projects.

Special Considerations
Procedures and software tools should comply with company format and computing capability.

Acceptance_____, Project Manager

Figure 9.10   Sample critical chain implementation project charter.

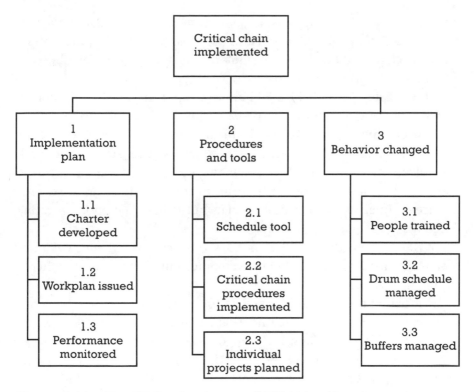

Figure 9.11 The WBS to implement CCPM identifies the work package deliverables.

- The drum manager creates the drum schedule to accommodate management's project priority.

- Project managers schedule projects to the drum schedule.

- Resources work to the roadrunner paradigm.

- Resources provide accurate input to the buffer report.

- Every manager does buffer management.

The work plan tasks developed following the WBS must create those results.

Consider the seven-S model and the actual behavior in your organization to complete your plan. Try to separate fact from fiction. For example, many people initially believe that all tasks in their organization are underestimated. That is often in an organization with extensive multitasking and interruptions. In some cases, they have data on actual reported task completion for previous projects. It usually requires only a quick check to find that they are similar to most organizations, showing extensive date-driven behavior.

The sample WBS in Figure 9.11 also considers the layers of resistance. Experience demonstrates that two-day critical chain training can effectively overcome layers 1, 2, and 3 of resistance. WBS element 3.1 thus includes the plan tasks to overcome those layers of resistance. The work plan elements supporting WBS element 1.2 must overcome layers 4 and 5. A work plan task to identify obstacles and potential negative consequences is one way to accomplish that, but it has a potential negative outcome.

WBS elements 2.2, 2.3, 3.2, and 3.3, once completed, give evidence that resistance layer 6 has been overcome, but they do not explicitly overcome layer 6.

Notice that the WBS in Figure 9.11 focuses on the required behavior change, not on changes in the underlying beliefs or culture. That accomplishes the most direct change, although it may not lead to lasting change. While some of the feedback from CCPM is self-reinforcing, it is legitimate, in many organizations, to fear that management's exploitation of the method will extend to exploitation in a negative sense, for example, over-loading the drum resource. If your organization is misaligned with the principles underlying TOC, you should take this opportunity to begin the culture and belief changes you need to continue with ongoing improvement. Otherwise, improvement will stop as soon as the implementation project ends.

The responsibility matrix shows responsibility for each of the work packages in the WBS. You can assign responsibility at multiple levels in the WBS, but you must assign it to the lowest, or work-package, level. Note that the person responsible for the work package is not necessarily the same as the resources required to perform the work contained in the work package. The responsible and accountable work package manager may be one of the resources that works on the project and may show up in their own work package and in other work packages. Work package managers plan and estimate the work package and then are accountable to manage its performance.

Figure 9.12 illustrates a schedule for implementing CCPM. It is in the critical chain format, as illustrated by the ProChain software–modified

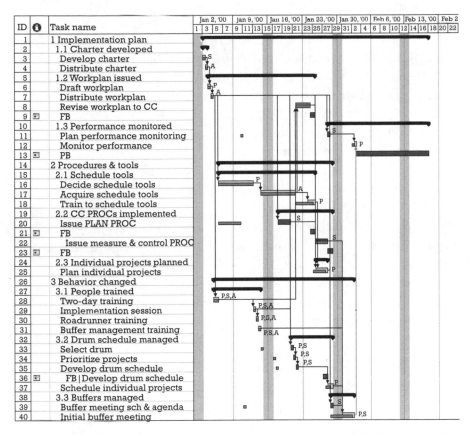

Figure 9.12 Critical chain plan to implement critical chain, shown in ProChain Gantt format.

Microsoft Project Gantt chart. The format illustrates all the tasks on the project but identifies the critical chain tasks by shade. Many of the plan bars are split horizontally, which supports showing status on the Gantt chart as the project progresses. It also shows the tasks that can be performed early without an adverse resource conflict on the project. Do not start those tasks early in a multiproject environment, because you may cause unnecessary schedule conflicts during project performance.

The letters adjacent to the bars indicate a specific type of resource. This chart shows only the end dates of the buffers. The plan, starting on January 2, 2000, predicts completion (end of the project buffer) on February 12. The project and feeding buffers are 50% of the preceding chain. Notice that the project plan does not develop this critical chain plan until about halfway through the project; it is the output of task 8, "revise work plan to critical chain." The plan assumes need for critical chain training, the software tool, and training on the software tool before completing the critical chain plan. You may choose to generate a critical chain plan from the outset. However, note that the critical path plan, which you would have used prior to the critical chain plan, has a later completion date without a buffer.

Note that the WBS elements include from one to four tasks. That is by no means a limit; work packages often have up to 25 tasks.

Because the complete critical chain plan makes it somewhat difficult to focus on the critical chain (especially if you cannot see the color difference), most project software allows you to filter the project tasks to display only the critical chain tasks. Figure 9.13 illustrates the critical chain for our implementation project. Notice that the critical chain includes both resource and path dependency. (The critical path plan previously referred to did not include leveling of the resources and therefore would have had resource conflicts.)

9.6.4 Plan to prevent or mitigate implementation risks

Project risk management seeks to control potential causes of special cause variation of high probability and consequence. The project plan monitors and may include prevention or mitigation planning for causes

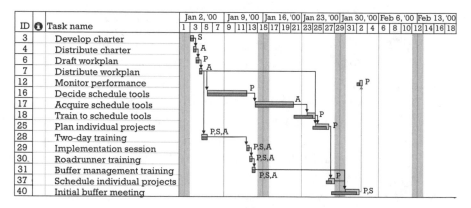

Figure 9.13 The implementation project critical chain identified by filtering the view for critical chain tasks illustrates both path and resource dependency.

of sufficiently high probability or consequence. Those special causes of variation are very organization specific.

Start by assessing the risk using whatever tools are appropriate to the magnitude of the project and comfortable to your team. Your plan can include a rudimentary level of risk assessment if your projects are small and pose no health, safety, or environmental risks. Chapter 10 provides an effective method for a wide range of projects and is appropriate for critical chain implementation. The other end of the risk assessment spectrum can include multimillion-dollar probabilistic risk assessments performed by teams of Ph.D.-level scientists, engineers, and legal and business experts. Critical chain implementation is at the low end of the risk spectrum. It does not affect the success of ongoing projects in a negative way, even if they do not achieve the critical chain benefits.

9.7 Move ahead!

Having completed all your planning, you are standing at the precipice of Goldratt's layer 6. At this point, the only way you are going to leave the airplane is to hold hands and jump. The following tables lay out the steps necessary to implement CCPM in a large, multiproject environment. Scale the steps to fit your needs.

Your team can accomplish all the actions listed in Table 9.2 in a one-day meeting. The meeting best follows a two-day workshop in which all the mangers learn the critical chain theory. It is a mistake to separate the two-day workshop from the implementation meeting, because people forget training rapidly if it is not immediately reinforced by application in the field.

Table 9.3 lists the steps necessary to begin implementation. Your leadership session may have identified the need for additional training. Deliver this training on an as-needed basis and do not let it delay proceeding with steps 4 and 5 of the process. You must perform steps 3, 4, and 5 with each individual project team. The planning sessions should not take more than a few weeks in total. If there are many projects, you may need to have multiple facilitators so this phase does not drag out.

Table 9.4 lists the steps necessary to complete the implementation phase and move into full-scale operation with the CCPM process. Buffer reporting should be initiated within three weeks of the initial leadership training, or the project will flounder. People will begin to forget what they

Table 9.2
Implement Phase 1

Step	Responsible Party	Action	Output
1	Facilitator	Plan a session with the leadership team and inform them of the agenda	Meeting schedule and agenda
2	Facilitator	Brief the team on the multiproject solution	Knowledge
3	Leadership	Leadership team identifies the constraint resource (drum) for the organization	Constraint (drum) resource identified
4	Facilitator	Present (briefly) buffer management	Knowledge
5	Leadership	Assign responsibility for buffer reporting	Responsibility assignment
6	Project managers	Commit to track and manage to buffers	Commitment
7	Leaders	Select initial project for CCPM	Project list
8	Leaders, project managers	Commit to duration for individual project CCPM plans and first buffer reports	Plan
9	Leaders	Commit to plan all future projects using CCPM	Commitment
10	Leaders	Determine project priority (or sequence for the drum resource)	Project priority list
11	Leaders	Assign responsibility to create the CCPM plans	Individual critical chain project plans
12	Leaders	Decide on the CCPM schedule tool	Schedule tool, procedure
13	Leaders	Identify who requires what training	Training matrix
14	Top leader	Commit to formally announce CCPM (duration)*	Commitment letter, e-mail, or meeting
15	Project and resource managers	Commit to communicate CCPM to people (duration)*	Individual communication
16	Drum resource manager	Commit to building the drum schedule (duration)*	Drum schedule
17	Project managers	Commit to weekly buffer meetings	Weekly buffer meetings
18	Facilitator	Get commitment for follow-up session	Follow-up session

*Insert the number of days estimated in the action.

learned about critical chain, and implementation success chances will dwindle.

Table 9.3
Implement Phase 2: Individual Project Critical Chain Plans

Step	Responsible Party	Action	Output
1	Facilitator	Deliver two-day workshops	Knowledge
2	Trainer	Train software users (if necessary)	Software skill
3	Project managers	Verify or create individual project plans suitable for critical chain plans, including normal (low-risk) task duration estimates	Individual project critical path plans
4	Resources	Determine average task durations	Input data to create plan (including buffer sizing)
5	Project managers	Create the individual critical chain plans, including sizing all buffers	Individual project critical chain plans (start dates not yet staggered)

Table 9.4
Implement Phase 3: Drum Schedule and Project Schedules

Step	Responsible Party	Action	Output
1	Drum manager	Create initial drum schedule	Drum schedule
2	Project managers	Schedule individual projects	Project schedules
3	Trainer	Train resources in roadrunner behavior and using buffer report to set their individual work priority	Knowledge
4	All project team members	Initiate resource buffer reporting and management	Buffer reports and action plans

Once you have moved into initial implementation, you will find a host of items that require clarification and issues that require resolution. You need an ongoing process to ensure that questions are answered promptly, answers communicated to all team members with a need (or desire) to know, and that you promptly resolve issues. This process can be part of your measurement and control process.

9.8   Measure and control implementation

Measurement and control of the CCPM implementation project provide the system feedback necessary to move your project management system to the new equilibrium state and keep it there. Your team must install a positive feedback loop to cause the change. Weekly buffer meetings are the primary vehicle for that feedback. They are your lever to lift the world.

Additional feedback during implementation includes the following:

▶ Prioritizing the projects sets a clear basis for decision making.

▶ The drum schedule and staggering of project starts using the drum schedule eliminate much of the serious resource contention for the project teams.

▶ Buffer management provides a clear decision-making tool to allocate resources between projects.

▶ Project resources are expected to work on one project task at a time and encouraged by management to protect this mode of operation.

▶ Project resources are not pulled away for higher priority projects.

▶ Project changes and subsequent rework are reduced due to later starts and earlier project completion.

▶ Project changes are reduced because the critical path does not change.

▶ Project changes are reduced because the buffer management thresholds for action are much wider than tolerances usually placed on project performance variation.

Management can enhance the effect of those natural feedback results by assuring communication throughout the project performance system.

Many natural feedback loops will help keep the CCPM system stable in its new state, including:

▶ Workers experience less stress (a positive feedback) when multi-tasking is removed.

▶ Project teams experience positive feedback from successfully completing projects.

▶ Management experiences positive feedback for increased project success.

▶ Management experiences positive feedback for increased profitability.

While some of the higher level feedback results start as soon as the projects begin to perform to the critical chain plan, and management begins to model the new behaviors, the feedback is relatively weak until the projects begin to complete.

Once management has planned and sequenced the projects, their primary roles are to:

- Participate in the buffer management process;

- Ensure that any new projects posed for inclusion into the system are prioritized and fit into the drum schedule.

9.9 What if implementation progress stalls?

Sometimes, implementation projects stall. That sometimes occurs near the beginning of the project but may occur at any time. People attend the training sessions and meetings and create the work plan. They then seem to completely forget about it in the cold light of the next Monday morning. They immediately drop back into the behavior patterns demanded and rewarded by the current system. Symptoms include complaints that "we are too busy to do critical chain" or that the software or procedures are not working right or that managers are not "walking their talk." You can expect one consistent symptom: No one blames himself or herself. That is resistance layer 6 in action.

If you have been working on the implementation for more than three months and are not essentially there and beginning to get the positive feedback, you are stuck. You need to dig into your organization's policies, measurements, and behavior to find out where you are stuck and implement a remedy to remove the block. I cannot give you a generic solution because all organizations are different. However, based on Dr. Deming's assertions that 96% of organization problems are caused by management (seemingly confirmed by my own observations), I can suggest that you start by looking at your leadership.

9.10 Summary

This chapter reviewed the theory of organization change and provided a practical plan for implementing critical chain in a multiproject organization. You can implement it on single projects with a simpler plan. Key points presented in this chapter are:

> Effective change plans harness organization dynamics (i.e., positive feedback loops) to accelerate the change.

> Senior management leadership is the critical success factor for multiproject CCPM implementation.

> Use your project process to implement the change to CCPM: charter, endorse, work plan, perform, and close.

> Pilot projects are best avoided for critical chain implementation.

> Humans and organizations try to maintain equilibrium (i.e., appear to resist change). Your implementation plan should anticipate and plan for this.

> Move ahead!

CCPM implementation is greatly aided by an organization's understanding of TOC and a successful change history of adopting new practices. Organizations also have succeeded the other way: using CCPM implementation as the catalyst for wider learning about TOC.

References

[1]   Dalziel, M. M., and S. C. Schoonover, *Changing Ways, A Practical Tool for Implementing Change Within Organizations*, New York: Amacom, 1988.

[2]   Covey, S. R., *Principle-Centered Leadership: Teaching People How to Fish*, Provo, Utah: Institute for Principle-Centered Leadership, 1990.

[3]   Bennis, W. G., K. D. Benne, and R. Chin, *The Planning of Change*, New York: Holt, Rinehart, and Winston, 1969.

[4]   Skinner, B. F., *Science and Human Behavior*, London: The Free Press, Collier Macmillan, 1953.

[5]   Culbert, S. A., *Mind-Set Management, The Heart of Leadership*, New York: Oxford University Press, 1996.

[6]   Argyris, C., and D. A. Schon, *Organizational Learning II, Theory Method, and Practice*, Reading, MA: Addison-Wesley, 1996.

CHAPTER

10

Contents

Project risk management

Project risk management seeks to manage and control the risk of project success to an acceptable level. Project risk deals with the risk to project success in terms of scope, cost, and schedule, including customer satisfaction. Other processes deal with other risks, such as health and safety risk and environmental risk. Project risk management seeks to control risks beyond the scope of your project plan and beyond your circle of control.

Project risk management is part of the project planning process, because you must decide on a course of action to include in your project plan based on the relative risk. Whenever you make a project assumption, you are making a project risk decision because you are assuming that reality in the future will follow your assumption. If your assumption does not come true, you have a project risk event.

Project managers have several options for dealing with project risk events, including:

▶ Expending effort to prevent the occurrence of the risk (e.g., limiting the use of flammable materials to prevent a potential fire);

▶ Identifying and monitoring risk triggers (e.g., weather monitors);

▶ Taking preventive actions that may reduce the potential consequences of the risk, should the event occur (e.g., spill control dikes);

▶ Purchasing insurance;

▶ Planning for mitigation in case a risk event occurs (e.g., fire department);

▶ Transferring the risk (e.g., subcontracting);

▶ Accepting the risk.

▶ Limiting the risk (e.g., setting allowances).

Critical chain simplifies conventional project risk management because it need deal only with special cause risks. The CCPM process provides the necessary and sufficient process and tools to deal with common cause risks to schedule and cost and to scope in some degree. The project quality process is also a risk management tool, protecting the project from scope risk.

The PMBOK™ [1], its more detailed support information [2], and much of the project literature (e.g., [1–4]) fail to discriminate between common cause variation and special cause variation when addressing project risk. We noted in Chapter 3 that Dr. W. Edwards Deming, the father of TQM, described that as a fatal error. The problem with not differentiating is that management will take action when they should not and not take action when they should. Deming notes a project example [5]:

Engineers in many establishments are allowed a deviation of 10 per cent between estimated cost of a project, and actual cost. The 10 per cent comes from stargazing: no basis whatsoever for it.... The control limits show that the natural variation of the differences on these 20 projects was 21 percent above and below the estimated cost.

10.1 Defining project risk management

Risk has two components: the probability of a risk event and the impact to the project.

Risk types include the following:

▶ Program risk: Risks that may cause client dissatisfaction, including risk that the client need is not known, that the full scope to fill the need is not known, or that project assumptions may not come true.

▶ Business risk: Impact the project may have on the rest of the business, including financial risk and risk to the company reputation.

▶ Cost risk: Potential to affect the project beyond one-third the project cost buffer.

▶ Schedule risk: Potential to affect the project beyond one-third of the project schedule buffer or beyond a feeding buffer.

▶ Health and safety risk: Potential for injury to the project team or public beyond the risks routinely accepted by the public.

▶ Environmental risk: Potential to affect the project necessary conditions (scope, schedule, cost) as a consequence of some environmental impact.

▶ Regulatory risk: Potential to affect the project necessary conditions (scope, schedule, cost) as a consequence of some regulatory impact, such as a new design requirement, constraint, or delay.

10.2 Risk management process

Figure 10.1 illustrates the project risk management process. It starts with identifying the risks your project may encounter.

Risk assessment may be quantitative or qualitative. Quantitative risk assessment tools include failure modes and effects analysis (FMEA), Monte Carlo analysis, project simulation, PERT, probabilistic safety assessments (PSA), and management oversight or risk tree (MORT). I focus on qualitative risk assessment because the data usually are not available to justify quantitative risk assessment; and supplying numbers tends to yield a false sense of believability.

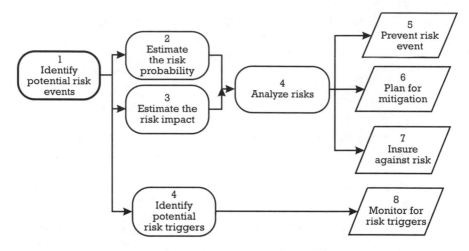

Figure 10.1   The project risk management process.

10.2.1   The risk matrix

Table 10.1 illustrates the basic risk management matrix. It summarizes the risk, the assessment of the risk, and the planned actions to monitor, prevent, or mitigate the results of the risk. The content in the table is for illustration only; your content should be much more specific to your project. However, I do encourage you to follow the lead of combining like risks to keep the overall length of the list to a reasonable number of items. Define what a reasonable number of items is based on the overall risk and size of your project. Relatively small projects (i.e., less than a few million dollars and under one year) should not have a risk list in excess of 10 items. If your list for a project of this size just seems to have many more high-impact, high-consequence risks, ask yourself if you really want to do that project!

The columns in Table 10.1 start with a description of the potential risk event. You can start with a list of many specific events that people can imagine and then lump them together for subsequent analysis. You can also categorize risks in terms of the probability and potential impact, the next two columns. That is, you may have one event for low-impact natural events and another for large-impact natural events. The reason to do that is that the two types of events will lead to different mitigation strategies.

The fifth column in Table 10.1 lists the triggers to monitor. They are the items you should frequently assess to see if you should change your

Table 10.1
The Risk Matrix

No.	Risk Event	Probability	Project Impact	Trigger to Monitor	Prevention Actions	Mitigation Actions
1	Natural event	Low	High	Weather reports, trends	Plan outside work during dry season	Use tents and tarps over work areas, dikes around facility, high-wind design, seismic design, pumps for rain and floods
2	Fire	Low	High	Fire prevention inspections; alarm system	Use noncombustibles, fireproof storage cabinets	Install fire suppression ASAP
3	Technical development	Medium	Medium	Development tests and gates	Institute quality process	Use alternative technology
4	Regulatory impact	Medium	Medium	Excessive questions or no action from regulators	Employ face-to-face discussions with regulators; use consultants to prepare applications	Form task team to respond to causes of delay or denial
5	Supplier delay	High	Medium	Late contracts; delayed delivery	Institute late-delivery penalties; check supplier delivery references	Use alternative suppliers, alternative equipment

risk assessment or activate your contingency plans. Of course, you should try to come up with leading indicators whenever possible.

Columns 6 and 7 of Table 10.1 are the most important: listing the actions you will take to prevent or mitigate the potential risk. Prevention and mitigation may work on either the event probability or the event impact. For example, a spill-control dike reduces the potential impact of a spill, but not the probability. On the other hand, a double-wall tank reduces the probability of a spill. Actions to prevent the risk should become part of your project work plan. Actions to mitigate may require actions in your project work plan to plan for mitigation, such as training or purchasing emergency supplies.

10.2.2 Incorporating risk assessment into the project process

Your risk assessment is only as valuable as what you do with it. Listing risks might give you ammunition to say, "I told you so." It also opens you to the question, "And why didn't you do anything about it?" You must take action on the identified risks to have any result from your risk analysis. You may choose to use a consistent approach to applying those alternatives, such as listed in Table 10.2.

10.3   Identifying risks

10.3.1   Risk list

You can use a variety of methods to identify risks. One method starts with the assumptions your team felt necessary to develop project work

Table 10.2
Guideline for Processing Potential Risk Events

		Probability of Risk		
		High	Medium	Low
Consequence of Risk	High	Prevent event; reduce consequences; plan to mitigate; monitor	Plan to mitigate; monitor	Plan to mitigate; monitor
	Medium	Prevent; plan to mitigate; monitor	Plan to mitigate; monitor	Monitor
	Low	Monitor	Monitor	Ignore

estimates. Each of the assumptions represents a risk of not being true. You can use checklists, such as those included in PMBOK™ [2]. You can use computer assistants, such as included in some software offerings [6]. You can simply get your project team together and brainstorm a list to start with. You can evaluate the problems encountered by previous similar projects. Coming up with the list is usually the easy part. You will never be able to predict the future, so you will never be able to come up with a complete project risk list. It would be infinitely long anyway and not very useful to your team. Instead, you should obtain a representative list of the types of risks likely to confront your specific project during its time of execution. Table 10.3 identifies categories of risk, and lists examples under each category. You can find checklist of potential risks in the project management literature.

Invite updates to the list as the project proceeds: additions, deletions, and modifications.

10.3.1.1 Project assumptions

Many of your project assumptions may translate to project risks if the assumptions do not come true. For example, your assumption that regulatory permit reviews will take 60 days low-risk and 30 days average duration may become a risk if the reviews take longer than two-thirds your project buffer. You may have reason to expect that the reviews could take much longer based on recent experience with the same regulatory agency on another project.

Table 10.3
Examples of Project Risk Events (After [2])

External Unpredictable	External Predictable	Internal Nontechnical
Regulatory	Market	Management
Natural hazards	Operational	Schedule
Sabotage	Financial	Cost
Technical	**Legal**	**General**
Technology changes	Licenses	First-of-a-kind
Design	Contracts	Remote site
Performance	Lawsuits	Foreign culture

On the other hand, you should guard against too many project assumptions. You need to ensure that a rule of reason applies in specifying both the project assumptions and the associated risks.

10.3.1.2 Checklists

Checklists often help to identify risks you might otherwise overlook, but they have two inherent problems:

▶ Checklists may suggest risks that are not significant to your project but that become believable once suggested.

▶ Checklists can lead to overconfidence that you have considered everything important, limiting your search for things beyond the checklist.

10.3.1.3 Plan scrutiny

You should scrutinize your plan by asking, "What could go wrong?" in each of the major steps to aid in developing your risk list. You can let the list get relatively long while you are preparing it; you will consolidate it in the next step.

10.3.1.4 Consolidation

If your risk list begins to get long, group like items to consolidate the risk list before going on to select the risk actions. Your purpose is to come up with a reasonable set of risk items to manage. Increasing the detail of the risk list does not increase its accuracy. There are an infinite number of potential risk events, so you can never list them all. It is far more important that you capture the important types of risks and put in place the appropriate information and response system to deal with the risks that do arise. You lose focus if the list of individual risk events becomes too long. It is impossible to plan realistic actions to prevent or mitigate the effects of too long a list. Try to limit the list to, at most, a few tens of items. For most reasonable size projects (i.e., less than about $10 million and one or two years in duration), the list should be less than 10 items (if not, you probably should not be doing the project).

10.3.2 Classifying risk probability

Make an estimate of the probability of each risk event actually occurring during the life of your project to decide on a rational plan to manage the risks. You do not want to spend a large amount of your project resources guarding against low-probability events. On the other hand, you want to prevent events that are likely to occur and be prepared to handle some events even if they are unlikely, if the potential consequence is large enough.

Bernstein notes, "The essence of risk management lies in maximizing the areas where we have some control over the outcome while minimizing the areas where we have absolutely no control over the outcome and the linkage between effect and cause is hidden from us" [7]. He goes on to note that insurance is available only when the law of large numbers is observed, that is, where the laws of chance work in favor of the insurance company or gambling institution. The very nature of risk, then, ensures that we are dealing with relatively low-probability events to start with.

People's ability to estimate probability is notoriously poor [8–10]. When considering probability, most people are subject to numerous logical biases and errors. Unfortunately, research demonstrates people are likely to be unjustifiably confident in their erroneous "knowledge." I will just list the common errors here to make you aware of them. Overcoming these biases and errors is the topic of another book.

- *Failure to understand how probabilities combine.* The probability of two independent events is the product of the probabilities of the individual events. Because probabilities are always numbers less than 1, the probability of the two events is always lower than the probability of either single event.

- *Failure to consider the base rate.* The base rate error fails to consider the distribution in the population. For example, consider a bead drawn from a population of 90% white beads and the probability of correctly identifying a white bead in dim light of 50%. A person looks at a bead under those conditions, and says "black bead." What is the probability that the bead was black? Most people answer 50%. The correct answer is only 5%.

- *Availability.* The availability error gives unjustified bias toward whatever comes to mind, usually because of recent reminder but also because it is something thought to be typical.

- *Failure to understand the law of large numbers.* People routinely accept small samples as indicative of a larger population and fail to understand that the variance in small samples tends to be much larger than the variance in larger samples or the population.

▶ *Representativeness error.* People mistake "more typical" for "more probable." For example, people will claim that a description of woman is more likely to be "a school teacher" than "a working woman," based on a description that includes traits people associate with schoolteachers. This answer is wrong because "working women" also includes more women than just women schoolteachers, therefore it is more likely that she would be a working woman than a schoolteacher.

▶ *Anchoring.* People tend to not deviate much from initial positions put forth by others or themselves, especially in regard to numbers. That bias also allows groups to significantly influence each other. If you want independent input, you have to seek independent input and not have one person review another's work.

▶ *Confirmation bias.* Once people have made a statement or a decision, they tend to look for instances that confirm that decision. Unfortunately, confirmatory cases have no value in scientific proof. People should look for instances that would disconfirm their hypothesis. That bias often results in worthless tests. Effective tests must always seek to disconfirm the hypothesis.

You can use that list to critically review your list of risk events and your categorization in terms of probability and impact. Ask, "Are we making this error?"

10.3.2.1 High probability

Do not have any items on your risk list that exceed a 50% probability of occurring during the life of your project. Count on items with a greater than 50% probability and include them in your project assumptions and baseline plan. Consider defining high risk as less than a 50% chance of happening during the life of the project but more than a moderate risk, which you might define as ranging from a 5% to 15% chance.

10.3.2.2 Moderate probability

The cop-out definition is that moderate-probability risk events are less than high probability and greater than low probability. They are events that may occur during the life of your project, but you would not bet on them happening (at least, you would want very favorable odds on the bet).

10.3.2.3 Low probability

Low-probability risks include risks unlikely to occur during the life of your project (i.e., less than a 5% chance of happening), down to a very low probability, perhaps on the order of 1% or less. Your project design may have to account for risks of lower probability during the life of the project result, such as earthquakes or extreme weather, but that is not the topic of project risk assessment. Exceptions may include insurance for events such as extreme weather (e.g., hurricane, fire, and flood) on a construction project.

10.3.3 Classifying risk impact

To define the risk, you must classify the risk impact because risk is the product of probability multiplied by impact. You could qualify impact in terms of the overall project schedule and cost or of the expected return on investment for the project. CCPM provides a unique measure of classifying the risk impact in terms of the project buffers for time and cost. The buffer size is an indicator of the common cause risk in the project and therefore is a reasonable basis to measure special cause variation. Sometimes fiduciary responsibility may require insurance for risks with chances greater than or equal to 1 in 1,000.

10.3.3.1 High impact

High impact is anything that could cause an impact in excess of the project buffer on schedule or in excess of the cost buffer on cost or otherwise result in client or project team dissatisfaction.

10.3.3.2 Moderate impact

Moderate impact is an impact that would consume on the order of two-thirds of your project buffers or all of your feeding buffers but at least one-third of the respective buffers.

10.3.3.3 Low impact

Low-impact consequences would not exceed one-third of your project schedule or cost buffers and would not be a significant concern to your client or project team.

10.4   Planning to control risks

10.4.1   Risk monitoring

Plan on monitoring for the risks you elect to keep in your risk management list. That means you should, at a minimum, review the list with the team members at your weekly project meetings and ask the question if any of the risk triggers seems to be imminent. Sometimes, you may need more formal monitoring for the risk triggers.

10.4.2   Prevention

Risk prevention activities you have elected to implement become part of your project plan. All you have to do to ensure that they are in place is to follow through on your project measurement and control process.

10.4.3   Mitigation planning

Plans for risk mitigation should also be part of your project plan. Include routine activities necessary to ensure the viability of your risk mitigation plans, such as fire inspections or emergency drills, as part of your project monitoring and control process. You need not include such periodic or ongoing activities as specific activities in your project network.

10.5   Summary

Project risk management controls special cause variation through monitoring, prevention, mitigation, or insurance.

- ▶ CCPM simplifies project risk management by eliminating the need to address common cause variation. CCPM risk management addresses only special causes of variation.

- ▶ You must include a risk management process in your project work plan. You should scale the implementation of risk management to overall project risk.

- ▶ Project risks must identify the risk event, the probability of the event occurring, and the potential impact or consequence of the risk event to the project.

▶ The CCPM project plan helps to define the relative risk in terms of the project buffers.

▶ The project team must decide among options, including prevention, mitigation, insurance, monitoring, and ignoring risks.

Risk monitoring, prevention, and preparations for mitigation should be part of your project plan. You should assign specific responsibility to monitor risks throughout the performance of the project and update your risk plans as appropriate.

References

[1] Duncan, W. R., et al., *A Guide to the Project Management Body of Knowledge*, Upper Darby, PA: Project Management Institute, 1996.

[2] Wideman, R. M., *Project and Program Risk Management, A Guide to Managing Project Risk and Opportunities*, Upper Darby, PA: Project Management Institute, 1992.

[3] Meredith, J. R. and S. J. Mantel, *Project Management, A Managerial Approach*, New York: Wiley, 1985, pp. 68–71.

[4] Wysocki, R. K., R. Beck, Jr., and D. B. Crane, *Effective Project Management*, New York: Wiley, 1995.

[5] Deming, W. E., *The New Economics*, Cambridge: MIT Press, 1993.

[6] RiskTrak™, Risk Services & Technology, Amherst, NH.

[7] Bernstein, P. L., *Against the Gods, The Remarkable Story of Risk*, New York: Wiley, 1996.

[8] Kahneman, D., P. Slovic, and A. Tversky, *Judgment Under Uncertainty: Heuristics and Biases*, Cambridge: Cambridge University Press, 1982.

[9] Belsky, G., and T. Gilovich, *Why Smart People Make Big Money Mistakes, And How to Correct Them*, New York: Simon & Schuster, 1999.

[10] Russo, J. E., and P. J. H. Schoemaker, *Decision Traps: The Ten Barriers to Brilliant Decision Making, and How to Overcome Them*, New York: Simon & Schuster, 1989.

CHAPTER 11

The TOC thinking process applied to project management

This chapter integrates the rest of this book and describes the process used to think through project management and develop the critical chain process. One of Dr. Deming's obstacles to improvement is a tendency on the part of many managers to "search for examples." He states, "My answer to such enquiries (i.e., for examples 'just like us') is that no number of examples of success or of failure in the improvement of quality and productivity would indicate to the enquirer what success his company would have" [1]. He notes further: "The question is not whether a business is successful, but why? And why was it not more successful? ... It is necessary to understand the theory of what one wishes to do or make." This text presented the CCPM theory you need, including some of the supporting TOC, TQM, and PMBOK™ principles. The Goldratt thinking process provides a tool to capture all of that reasoning. It is up to you to put it to work.

11.1   Applying Goldratt's thinking process to project management

The thinking process, described in Chapter 2, is Goldratt's attempt to enable the rest of us to create the innovative system solutions that he developed. The process starts with a model of current reality and ends with a plan to create a preferred future reality. It is a logical process that depends heavily on critical review and buy-in to the models.

In seeming contrast to Deming's assertion, I claimed in Chapter 2 that you do not need to understand TOC tools to successfully apply CCPM. The conflict cloud evaporates by understanding the underlying assumptions of CCPM expressed throughout this book. The understanding of "why it is successful" may be sufficient for your application. Comparing the assumptions identified in this book to your organization's reality may enable you to adapt the process as necessary. Certain features of the process (e.g., eliminating bad multitasking) appear to be so robust that you are likely to experience positive results with CCPM even if your implementation does not match the textbook representation provided here. Organizations that are relatively flexible (which some call "early adopters") have succeeded in deploying critical chain without knowing the thinking process and with only an overview knowledge of TOC tools.

I had the fortunate experience to see an early version (August 1994) of the thinking process applied to project management developed by Dee Jacob of the Avraham Goldratt Institute (AGI). It included the thinking of Jacob, Dr. Goldratt, and others at the Institute. It had all the essentials presented in current thinking for single-project critical chain. It made sense to me in view of my years of project experience. The changes from 1994 to the current process are minor in substance and have more to do with displaying the results than they have to do with the essential project management system.

The initial method did not include the current understanding of multiproject critical chain. Dr. Goldratt had devoted only a couple of pages to multiple-project resource contention in *Critical Chain*, published in 1997 [2]. From my view, the significance of the multiproject solution was really discovered and developed by Tony Rizzo of Lucent Technologies, who has not yet published it on his own. In that regard, the result seemed to actually lead the process to develop it.

Unfortunately, few people seem to have the inclination to engage in the deep thinking and scrutiny necessary to follow the thinking process. Noreen, Smith, and Mackay describe some early results with training in the thinking process [3]. Their results show poor application after the early Jonah training. (Jonah training is the name that AGI uses for training in the thinking process, following the behavior of the character Jonah in Goldratt's first book, *The Goal* [4].) My (nonscientific) poll over the last several years leads me to believe that those findings would not change substantially today. That is, few people apply the thinking process after the Jonah training, and fewer yet create breakthrough solutions. There are even some reasons to believe that creative solutions never proceed in the inherently inductive path implied by the thinking process [5, 6]. Popper and DeBono might agree that people like Goldratt, in reality, leap to a new hypothesis and then use the thinking process to justify it. Many people have followed through to develop the thinking process representations for TOC generic solutions, such as production and project management. The trees provide better and better understanding of those systems and may aid further improvement. They may also inhibit further improvement, because they tend to put boundaries around the problem.

Understanding the thinking behind critical chain may help you discern if your system conforms to the assumptions about current reality made in the process to develop critical chain. If your system differs in substantial ways, you may have to modify the future system and the injections to get a future reality that works for you.

The trees created as part of the thinking process are relatively complex and thus not appropriate for publication in book format. You can find them on the Advanced Projects Institute Internet site: www.advanced-projects.com.

11.2 Current reality tree

The CRT describes the system as it is today to help find the core conflict. The core conflict is a root cause of many UDEs. You start the CRT process with UDEs, which are those things that really bother you about the current reality. For example, "It really bothers me that projects overrun the schedule." (Chapter 3 described UDEs for project management.)

You then select three of the UDEs to develop a core conflict. You do that by developing each UDE conflict and then combining the three conflicts to discern the underlying generic conflict that leads to all three. Chapter 3 also illustrated the combination of evaporating clouds for project management. The CRT drives the cause to a core conflict that leads to most (and usually all) of the UDEs. A core conflict is not *the* core conflict. It is an important conflict and therefore a good high leverage place to focus on changing the system.

Figure 11.1 illustrates the base of the project system CRT, containing the core conflict developed in Chapter 3 (see Figure 3.9). It illustrates the cloud in sufficiency tree format, highlighting the assumptions that lead to the conflict. Reading from the bottom, "If everyone wants projects to succeed, and if increasing competition drives managers and clients to demand projects to get the most scope for the least cost within the shortest schedule, then successful projects must deliver increased scope and reduced cost and schedule." Continuing up the tree, "If successful projects must deliver increased scope and reduced cost and schedule, and if the only way to reduce the schedule of critical path plans is to reduce the duration of tasks on the critical path, and the only way to reduce cost is to reduce task cost, then there is pressure to reduce each task estimate."

You can read the right side of Figure 11.1 up to entity 9, which states, "There is pressure to include contingency in each task estimate." Including contingency conflicts with reducing task estimates, thus leading to the fights that surround project planning.

Note that I left out the "more and more" statements. They come about as you traverse the tree over and over in the same organization. Note where some of the UDEs, which are conclusions farther up in the tree, feed into the entities at the base of the tree, such as UDE-1, 2, and 3 feeding all the way down to entity 9.

Figure 3.10 illustrated the notional connection of the UDEs that flow from the base of the tree. The logic of the tree is not evident at this summary level. The actual tree includes many steps of intervening logic that describe how organizational beliefs and actions lead from one UDE to the next. The key point of Figure 3.10 is that all the UDEs in the tree are causally related and derive from the core conflict.

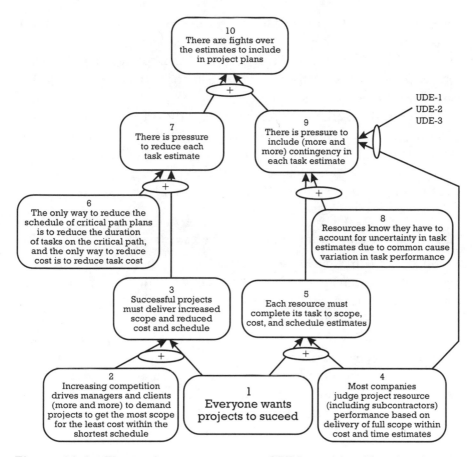

Figure 11.1 The project management CRT base identifies the core conflict.

The generic project management CRT cannot represent your environment. I have worked with companies that have started from very mature project management systems as well as with companies with a simplistic approach to project planning and control. Interestingly, it seems that organizations with the simplistic approaches are more able to adapt to CCPM. (Some argue that any degree of discipline would have helped as much. I dispute that claim because these are all highly multi-tasked environments, and conventional project management would only exacerbate that problem.) It is not unusual to find people whose idea of a project plan is a Gantt chart with no task logic, and only a final due date demanded by someone outside the project organization. In a few months, it is possible to lead those people to have effective critical chain project plans, buffer management, and a growing sense of success.

On the other hand, I have worked with organizations that have very detailed project plans (with thousands of activities) that, after six months, are still unable to resource-load their plans. In one case, the organization neglected to address the fact that the parts of the organization necessary to develop the complete project reported (collectively) to 13 different vice presidents in the organization, 12 of whom were not bought in to the changes necessary to implement CCPM. Guess which VP was replaced three months into the process. I think you can predict the result.

11.2.1 Policies, measures, and behavior

It is often useful to explicitly consider the impacts that policies and measures have on behavior and how that behavior affects current reality. Policies and measures are often the constraint of systems like project management. For example, entity 4 of the CRT, "Most companies judge project resource performance based on delivery of full scope within cost and time estimates" may be reflected by performance appraisal policies and measures. Very often, companies have efficiency measures and policies that reinforce multitasking. You should consider your company's policies that may reinforce the generic CRT.

11.2.2 Feedback loops

The detailed CRT logic contains feedback loops. For example, Figure 11.1 illustrated UDEs 1, 2, and 3 feeding into entity 9 at the base of the CRT. That circular nature of real systems bothers some people, who think in terms of cause-effect reasoning and want to know, "Which comes first, the chicken or the egg?" In the real world of dynamic systems, it is an ongoing circular pattern that evolves over time. That is one reason that people looking for root cause often get the wrong answer; the cause is the overall structure of the system, not an individual entity. All variables in the system change in a correlated way (although often with time delays).

The feedback loops are often where you can find the most leverage to change the system and thus are worth considering explicitly. Your measurement systems almost always constitute a feedback loop (if they do not, why are you making the measurement?). The primary feedback key to CCPM is management's reaction to task performance relative to estimated 50-50 durations. Management's response to buffer reporting is also critical.

11.2.3 Scrutiny

Scrutiny critically reviews the products of the thinking process. It is the primary way to determine if the hypothesis of the thinking process connects with reality better than alternative processes. (The secondary method is the tried-and-true method of scientific experimentation, something that is difficult to do with systems as complex as projects.) Chapter 2 described how, using the scientific method known as the TOC theory of knowledge, you can never prove the truth of any assertion or hypothesis. The best you can hope to do is, through critical review, assure yourself that one hypothesis describes reality better than another.

Scrutiny provides the thinking process critical review. You do it by subjecting each product of the thinking process to the categories of legitimate reservation. I learned these categories from the partners and associates at AGI, but I suspect they have some prior source in logic. Unfortunately, Dr. Goldratt does not use references in his books, and AGI does not reference the source of much of their material. Dettmer [7] is the only source I am aware of who has published these categories, but he refers to AGI as the source.

The categories of legitimate reservation are these:

▶ *Clarity*. Does everyone understand the meaning of the words in the entity?

▶ *Entity existence*. Is there evidence to support or refute the existence of the entity in reality?

▶ *Causality existence*. Is there evidence to support the claim of causality? (This evidence is usually of the form that the effect always exists when the cause is present and never exists when the cause is not present.)

▶ *Cause insufficiency*. Do other entities also have to be present to cause the stated effect?

▶ *Additional cause*. Can the effect exist without the stated causes but instead in the presence of other causes? (The good enough TOC principle suggests you limit your model to causes that represent at least one-third of the effect instances.)

▶ *"House on fire" (or cause-effect reversal).* "If there is smoke, then there is fire" is a common expression that does not reflect causality, even though it is expressed as "If-then." The accurate statement is, "If I see smoke, then I know that the house is on fire." The smoke, however, does not cause the house to be on fire.

▶ *Predicted effect.* Would the existence of the entity also cause some other effect? Does that other effect exist in reality?

▶ *Tautology.* Ayn Rand's famous "A is A." Or, as the story goes,
Joe: "Blowing a horn every day keeps elephants out of my living room."
Dan: "How do you know that?"
Joe: "Do you see any elephants in my living room?"
Detecting the logic error is sometimes hard. How many people call those psychic hotlines?

Jonahs subject each thinking process tree to critical review for compliance with those criteria. If you think getting people to think through the trees is difficult, try going through the review process on each entity and causality in the tree. Most people find it agonizing. Unfortunately, there does not appear to be a better alternative.

11.2.4 Buy-in

The people who will deploy the results of the thinking process have to agree with it to the extent that they are willing to make or tolerate (depending on their location and effect on the system) the changes it requires to move to future reality. Unfortunately, experience demonstrates that logic rarely influences people's beliefs. As Chapter 9 described, if you do not change beliefs, you are unlikely to succeed in changing behavior for the long term. You may cause a temporary effect, but the system will, over time, swing back to where it was before you started.

Buy-in seeks to achieve sufficient agreement to make the injections necessary for future reality. You will not achieve complete changes in belief at the outset, but if the future reality contains sufficient positive feedback to meet the real needs of the people who influence the system, the feedback will take over driving the system. You can achieve that with some people by leading them through the thinking process. Experience, however, demonstrates that it is the rare senior manager who will focus long enough to follow that through. Thus, you will need to use specially designed tools, presentations, and dialog to reach those people.

I have found that most people intuitively appreciate the logic and use of the evaporating cloud. You can even get senior managers to listen to the three clouds that lead to CCPM and to buy into the core conflict and then to the necessary injections.

A powerful method to enhance buy-in is to first solicit the UDEs of those you need to agree and include one or two of their UDEs into your CRT. Demonstrating how the system causes their UDEs usually succeeds in bringing about buy-in, not only to the CRT but also to the injections you propose to create future reality.

11.3   Future reality tree

The FRT describes what to change to. It is your vision of the future. Future reality does not exist when you start to make the changes to eliminate the UDEs of current reality. The outputs of the FRT are the injections we have to create for future reality to exist. Injections are effects that, if in place, will cause future reality. Injections are not (generally) actions; you develop actions as part of how to cause the change.

11.3.1   Desired effects

Start the FRT by changing each UDE into its opposite desired effect (DE). Figure 11.2 illustrates the map of DEs for a successful project management system, including an indication of where the injections tie in.

11.3.2   Injections

Injections are the changes you will make to the system. The FRT connects your injections to the desired effects of future reality. You then methodically work through the tree, determining where injections are necessary to create the future reality. Developing effective injections is the most creative stage of the thinking process. Experienced TOC experts like to word injections as completed effects, knowing that they will often have

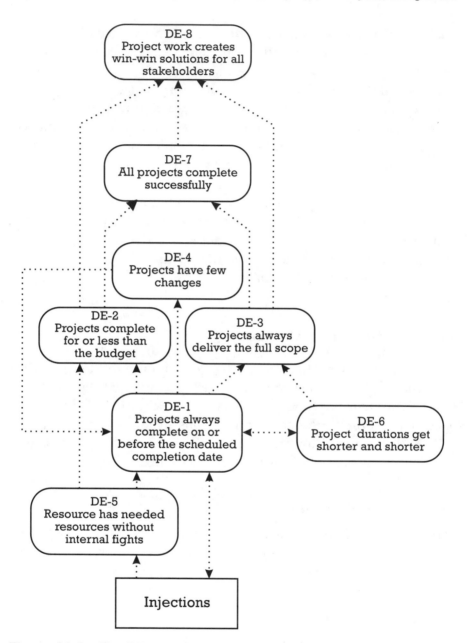

Figure 11.2   The DE map illustrates the relationship of the DEs in the FRT.

to develop a number of actions to achieve the injections. Collectively, the injections will create a future reality in which all the DEs exist.

The total list of injections for the single project are as follows.

▶ Reduce duration estimates to 50-50 estimates. Project managers identify the project's network of activities and paths by unbuffered time and by resource. Activity durations are normal estimates, which we know to be high probability.

▶ Eliminate resource contentions and identify the critical chain. Project managers identify the critical chain as the longest chain of dependent events, including resolving resource contentions. This is the first focusing step.

▶ Insert a project buffer sized and placed to aggregate critical chain contingency time (initially 50% of the critical chain path length). This is one step to exploit the constraint.

▶ Protect the critical chain from resource unavailability by resource buffers. Project managers correctly place resource buffers to ensure the arrival of critical chain resources. This is a second step to exploit the constraint.

▶ Size and place feeding buffers on all paths that feed the critical chain. Project managers use the feeding buffers to immunize the critical chain from accumulation of negative variations on the feeding chains. This subordinates the other project paths to the constraint.

▶ The plan schedules activities to start as late as possible, protected by buffers. This injection helps to further subordinate the other paths to the constraint by allowing the critical chain (usually) to start first, with at most a few other paths.

▶ Resources deliver roadrunner performance (eliminate multitasking and the student syndrome). The resources work as quickly as possible and as soon as possible on their activities and pass their work on as soon as they complete. This injection begins to elevate the constraint.

▶ The project manager provides resources with activity durations and estimated start times, not milestones. This injection helps to break the current win-lose paradigm associated with getting work done by the milestone date. It aids in encouraging resources to pass on their work when done. It aids in elevating the constraint.

▶ The project manager uses buffer management to control and plan. The project and feeding buffers provide the information to the project manager when to plan for recovery and when to take recovery actions. It also aids in elevating the constraint.

Figure 11.3 illustrates the injections in the sequence they must occur. Numerous logical steps supplement the injections to lead to the DEs.

11.3.3 Future reality tree

The FRT becomes the guide for change. As you implement injections on the FRT, you use it to monitor if you are achieving the DEs. You can also use the FRT as a resource to derive unintended consequences of the proposed changes. Note that changes that occur faster or larger than predicted by your FRT are also a cause for reassessment, since you may have missed some feature of the causalities.

The FRT provides a check on the CRT. You will often discover additional causalities that exist in current reality as you develop the FRT. Because the FRT focuses on the future, it is not necessary to go back and revise the CRT.

The bottom of the FRT starts with the injections summarized earlier. Each injection includes adjacent entities that describe why we need the injection and the logic explaining how the injection satisfies the need.

11.3.4 Feedback loops

The FRT contains the feedback loops to move the system to the new state and keep it stable. The project FRT exhibits a number of feedback loops, some of them with short delays and some of them with longer delays. The short-delay loops help to establish the system in the first place, and the long-delay feedback loops help to keep it stable.

One feedback loop illustrates the control effect of buffer management. When the project manager acts to restore a buffer to less than two-thirds penetration, that will cause some activities to perform in a shorter time.

Another feedback loop shows that as teams build confidence by completing projects successfully with CCPM, they act to further reduce overall planned lead time.

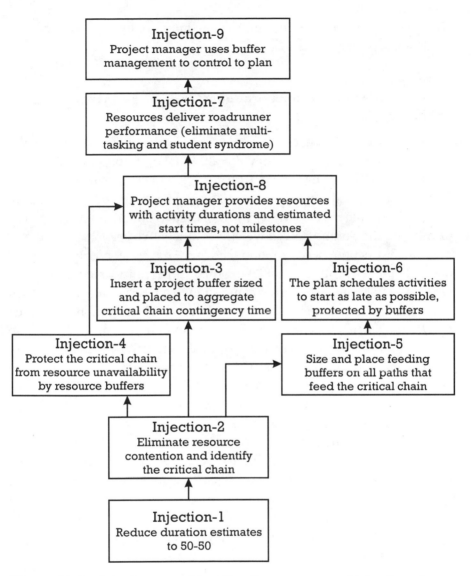

Figure 11.3 Injection map identifies the sequence of injections for the FRT.

Experience indicates that the project feedback loop activates long before the first project completes, leading to further reductions in the initial project lead time. That is likely due to increased confidence as the first part of the project shows little difficulty performing to the critical chain schedule. The project manager contributes by not criticizing those who do overrun reduced duration task times, as long as they exhibit roadrunner performance.

11.3.5 Unintended consequences (a.k.a. negative branches)

Unintended consequences are undesired results that occur from the actions we take. In a sense, they are an equal and opposite reaction to whatever action we have taken on the system. Sometimes (rarely, it seems), unintended consequences can be good. Often, however, they are not good. For that reason, we call the tool used to understand and prevent or mitigate the potential negative effects a negative branch.

The distinction between obstacles and negative branches is that obstacles prevent you from achieving the future reality or ambitious goal you have set. Negative branches come about because you have succeeded in creating the injection you intended to create. The injection, combined with other factors in current reality (or, sometimes, new factors also created in future reality), conspire to cause a negative outcome.

The major resource for identifying potential negative branches is the people who review your FRT. They have the intuition to understand how the changes you are going to create may interact with their reality to create an unintended consequence.

Figure 11.4 illustrates a potential negative branch dealing with the commonly voiced concern that if you make the safety time in the schedule evident, people will want to cut it. People usually include both customers and management that is more senior. The project management literature addresses that concern, often with a caution to "keep your safety time hidden." That hardly seems like a professional way to run a business!

You first build the tree to connect from the injections expected to cause the undesired effect to the stated undesired effect. The negative branch is a sufficiency tree, just like the CRT and FRT, and is read "If-then." By this time, we trust you have sufficient comfort with the construction to read the tree. You must check the tree to ensure that the entities and causalities exist and that the logic is complete and sufficient.

The next step in using the negative branch is to find an injection that will prevent the negative effect. You do that by examining the branch and locating the point at which it turns negative. In this case, entity 602, "The customer wants lead time reduced by cutting buffer times," is

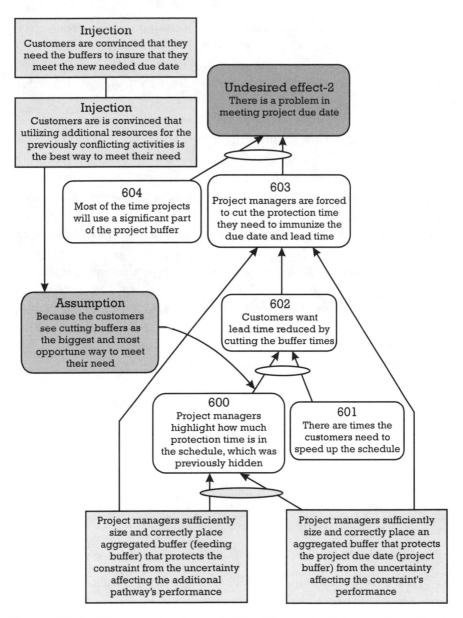

Figure 11.4 The negative branch identifies potential negative effects from the changes we make. That enables us to prevent unintended consequences.

where this branch turns negative. You then assess potential assumptions under the causality arrows that feed that entity. In this case, since only one causality feeds entity 602, you have to look at only two arrows. Note that since the connections from 600 and 601 to 602 include an *and*, eliminating either 600 or 601 results in eliminating the effect, 602. That is the meaning of the *and*. You need both feeding entities for the effect to follow.

In this instance, entity 601 appears to be a fact of life, so there would be little advantage to questioning the assumptions surrounding its existence or causality. There are assumptions in the causality between 600 and 602, the most obvious one, as noted on the tree, "Because the customer sees cutting buffers as the biggest and most opportune way to meet their need." That is likely to be a true assumption when the customer (which may be internal management) does not understand the ideas behind CCPM.

Once you have an assumption, you can propose alternative injections that make the assumption no longer correct or applicable. Two are presented on the figure. Either injection should do the job of eliminating the assumption and therefore preventing the UDE of this negative branch. Take your choice.

The negative branch procedure follows:

1. Identify the potential undesired effect of concern.

2. Identify the injection you suspect leads to the undesired effect.

3. Build a sufficiency tree to connect logically the injection to the UDE.

4. Scrutinize the logic in the tree (branch) by reading it aloud to others and having them agree to the logic.

5. Determine where the branch first turns negative.

6. Expose the assumptions under the arrows feeding the first negative entity on the branch.

7. Identify injections that will invalidate the assumptions and therefore prevent the negative effect.

It is, of course, possible that the injections you propose to trim the negative branch may themselves lead to unintended consequences. If so, examine the new negative branches before completing the strategy.

An injection is not a plan. It is not even a coherent strategy. Goldratt suggests the following tools for such purposes.

11.4 Prerequisite tree

The PRT is a time-phased tree of the effects that we must cause for the FRT to result. You assess each injection on the FRT to determine the obstacles that must be overcome for the injection to exist. You create intermediate objectives (IOs) to overcome the obstacles and logically link them in a time sequence with the injections. Figure 11.5 illustrates the PRT for the first critical chain injection. The obstacles are shown in the hexagons.

You read the PRT from the top down as follows: "In order to reduce duration estimates to 50% probable estimates, we must have people understand that they are expected to achieve only the shorter duration 50% of the time and that the feeding and project buffers protect the project, because most people feel that their duration estimates are already too short." Although some of the statements may get a little long, this representation provides a coherent sequence for the changes and the basis for the overall sequence. The phrases "In order to," "we must," and "because" connect each set of blocks on the tree. Reading the tree out loud is a good way to check the logic.

Add each new injection to the overall PRT to accomplish future reality. Each injection must have a transition tree (TRT) to implement the actions necessary to achieve the injections and/or IOs that build to the injections.

The PRT also has standalone utility as a tool to plan and achieve ambitious goals. Dr. Goldratt illustrates its use for that purpose in *Its Not Luck* [8]. It has great power to get a team to identify all the obstacles they foresee at the beginning of a project and to create a plan based on overcoming all those obstacles.

You have to use caution in developing the PRT obstacles, though, because a team may tend to create false obstacles in the face of future change that may affect them. You have to ensure that they have bought in to creating the objective of the PRT before you solicit obstacles. I have found that on CCPM implementation, people rarely follow through on PRTs created early in the process. They quickly find that many of the

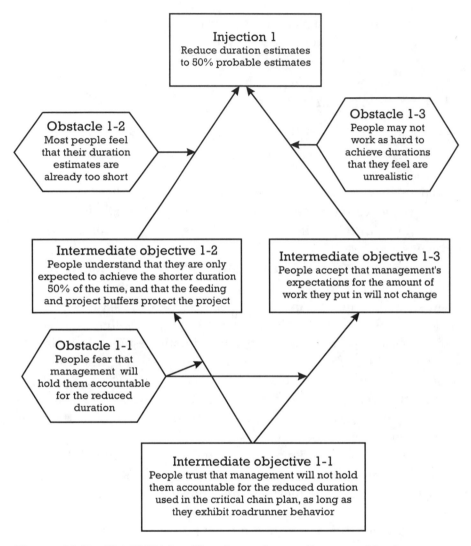

Figure 11.5 The TRT identifies the actions, effects, and logic to achieve the IOs. It provides clear instructions.

perceived obstacles do not exist. They also find that the real obstacles did not get on the PRT, sometimes because the organization culture and attendees in the session did not make it safe to discuss the real obstacles. It is much more effective to go on with implementation planning as presented in Chapter 9 and deal with the real obstacles as they arise.

11.5   Transition tree

The TRT provides the time-phased action plan to achieve the effects on the PRT. The TRT ties the actions to the logic for doing them and provides clear instructions to those who perform the activities. You can also use it to measure progress in terms of the EFFECTS produced, not the perform-ance of the action. The TRT has broad application for achieving any effect you wish. (For example, you can use TRTs to get buy-in to the thinking process results.)

Figure 11.6 illustrates a TRT for one of the project management system injections. The tree creates the first IO on the PRT in Figure 11.5. Create the TRT for only two levels of the PRT at a time, starting from the bottom. As you complete the IOs at one level, create the TRT for the next higher level.

You read the TRT from the bottom up, the same way you read the CRT and the FRT. For example, starting on the bottom of Figure 11.6: "If people will first look for inconsistencies in the written company reward system, then the written company reward system must align with critical chain behavior. If the written company rewards system must align with critical chain behavior, and if we revise company policies to reward 50% estimates and roadrunner behavior, and if when people see it in writing, they will suspect that a real change is possible, then policies support criti-cal chain behavior of 50% estimates and roadrunner behavior." The TRT describes the logic for each action and why we expect the action to create the desired effect. Some argue that the logic is so obvious, it need not be written down. Often what is obvious to one person is not obvious to another. The TRT has proved to be an effective tool for communicating clear instructions. It reflects why you are taking the actions you take.

The TRT also has standalone utility as a way to present procedures.

11.6   The multiproject process

11.6.1   Multiproject current reality tree additions

The multiproject environment adds the following additional UDEs:

 ◗ Management commits to project dates that are unachievable.

 ◗ Project managers fight over resources.

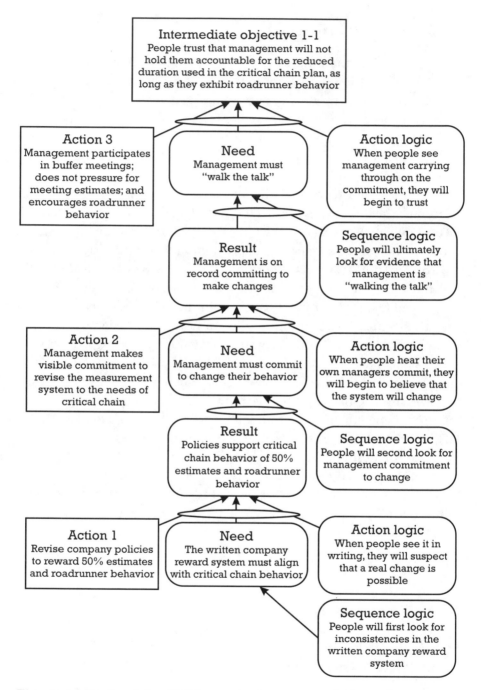

Figure 11.6 Read the PRT for project management from the top down.

- People have to work on multiple tasks for multiple projects.

- People are pressured by project managers to complete that project manager's task first.

- People are forced to work extensive and uncompensated overtime.

You should not find it difficult to extend the logic from the single project CRT to derive these effects, if they exist in your environment. The base of the CRT (core conflict) need not change.

11.6.2　Multiproject future reality tree additions

The multiproject FRT requires adding the following injections:

1.　Identify the drum resource for the organization.

2.　Identify the priority of projects for scheduling of the drum resource.

3.　Develop project start and completion dates from the individual project plans and the drum resource schedule.

4.　Insert a capacity constraint buffer between the use of the drum resource on projects.

5.　Insert a drum buffer upstream of the use of the drum resource in each project.

The multiple project FRT requires all the logic and injections from the single project FRT to feed FRT injections 3 and 5.

11.6.3　Multiproject prerequisite tree additions

The multiproject PRT therefore requires adding the following IOs:

1.　The drum resource is selected.

2.　The drum manager is identified.

3.　Initial project priorities are assigned.

4. Management commits to new project completion dates only after deciding the new project priority (relative to ongoing projects), developing the new project plan, and scheduling through the drum resource.

The implementation process presented in Chapter 9 includes those PRT requirements.

11.7 Future directions

TOC and TQM provide the tools for ongoing improvement to your project management process and to your entire enterprise. The TOC five focusing steps provide a strategy to continue improvement on a system. In addition, you can expand the definition of the system to widen the impact. For example, the analysis to develop CCPM left out significant parts of the PMBOK™. The remaining parts can be improved as well, including human resource management and project quality. Those areas are amenable to the thinking process to discover the core conflict and provide a plan for effective implementation.

Another direction for improvement is to look into the process steps in more depth, in particular the processes that create the technical plan and the processes that create the task results. You can view them as production processes or as projects.

I personally feel that the greatest improvements lie in improving management processes. I see a continuing trend toward more and more ineffective management as people inappropriately apply technological tools such as pagers, cell phones, and e-mail. More and more interruptions mean less and less focus, leading to a frenzied pace and little accomplishment. I see the image of a fibrillating heart thrashing away while producing no output. Many managers I meet with today have their full workday scheduled for them by others' meetings. If they cannot manage themselves, how can they expect to effectively lead others? Most of the workers in their groups report increasing multitasking, longer and longer work weeks, and more and more pressure.

The real promise of TOC lies in the impact on people who work in TOC organizations. As in some of the successful TQM organizations, people report a sense of order and hope. People report greatly increased work output with much less time spent and greatly reduced frustration. More and more, TOC organizations translate these gains into the bottom line for all the business stakeholders.

11.8   Summary

The thinking process provides the integration tool to expose the why and how to improve project management. It clarifies the beliefs and assumptions behind the theory, making them accessible for future improvement. Key points from the TOC thinking process application to project management are:

- The core conflict, leading to all project UDEs, is between the individuals and the project management system.

- The core conflict derives from how the project system manages (or fails to manage) uncertainty by allocating contingency.

- The constraint for single projects is the critical chain, the longest path through the project considering both the project task logic and the resource constraint.

- A system to exploit that constraint aggregates individual task uncertainty into buffers at the end of activity chains.

- Buffer management provides a real-time information (the answer to the question asked) system to effectively manage projects to complete on or before the scheduled end of the project buffer.

- The system constraint in a multiproject environment is a resource shared across multiple projects (the drum resource).

- Exploiting the multiproject resource constraint requires eliminating bad multitasking of all project resources.

- You must subordinate resource efficiency measures to the multiproject constraint. (Note that all resources other than the drum must have excess capacity.)

You should modify the model as necessary to match the specific requirements of your environment. So far, the generic injections have proved to be robust over a wide range of project environments. Specific deployment (e.g., implementation plan, PRT, TRT) varies. Experience to date demonstrates that effective leadership and a good implementation plan are the critical factors to successfully implement CCPM.

11.9 Closure

I do not contend that CCPM is The Next Big Thing (TNBT) [9] in project management, much less for management in general. As I noted from the beginning, it is not even the complete project management system; you need to have all parts identified through the PMBOK^{TM}. CCPM provides a fresh look at the project system with a number of elements that have survived both critical review and field experiments, compared with the baseline critical path method. In the words of many TOC experts, it has proved "good enough" to substantially improve project performance on many types of projects in many environments. Perhaps most significant, the people working in the CCPM environment consistently report reduced stress and increased joy in work, even though they are producing more.

So far, all organizations in which the management has truly committed to deploy CCPM have succeeded in getting substantial results, far exceeding the effort expended to make the change. It does not take a lot of management time, usually three days up front to learn and plan the change and a few minutes a week to review and act on buffer reports. Results should be visible in three months. The management time investment pays back rapidly through reduced need for resource battles, project changes, and expediting.

Organizations that try to just train the team or perform a single-project pilot in a multiproject organization are less successful. They frequently see little impact after six months. They have convincing explanations as to why they have not yet converted to critical chain plans and why they are not issuing and acting on buffer reports. The reason for this bi-modal result is that CCPM requires some significant (even if simple) changes in management behavior. You will achieve the benefits of CCPM only if management behavior follows the CCPM model.

It is a delightful experience to go to TOC meetings and hear people report their accomplishments with pride and a sense of relief. The only lament I have heard from people in TOC organizations is one I repeat myself, "I wish had learned this 25 years ago." I close with this question: "Can you afford to have your competitors successfully deploy CCPM before you do?"

References

[1]   Deming, W. E., *Out of the Crisis*, Cambridge: MIT Press, 1982.

[2]   Goldratt, E. M., *Critical Chain*, Great Barrington, MA: North River Press, 1997.

[3]   Noreen, E., D. Smith, and J. T. Mackey, *The Theory of Constraints and Its Implications for Management Accounting*, Great Barrington, MA: North River Press, 1995.

[4]   Goldratt, E. M., *The Goal*, Croton-on-Hudson, NY: North River Press, 1984.

[5]   Popper, K., *Objective Knowledge, An Evolutionary Approach*, Oxford University Press, 1972.

[6]   DeBono, E., *Lateral Thinking, Creativity Step by Step*, New York: Harper & Row, 1970.

[7]   Dettmer, H. W., *Eliyahu M. Goldratt's The Theory of Constraints, A Systems Approach to Continuous Improvement*, University Bookstore, 1995 (now available through ASQC Press).

[8]   Goldratt, E. M., *Its Not Luck*, Great Barrington, MA: North River Press, 1994.

[9]   Pinto, J. K., "Some Constraints on the Theory of Constraints: Taking a Critical Look at the Critical Chain," Project Management Institute, *PM Network*, August 1999, pp. 49–51.

List of acronyms and abbreviations

ACWP   actual cost of the work performed

BCWS   budgeted cost of work scheduled

BCWP   budgeted cost of work performed

C/SCSC   cost/schedule control system criteria

CCB   capacity constraint buffer

CCFB   critical chain feeding buffer

CCPM   critical chain project management

CPM   critical path method

CRT   current reality tree

CSCS   cost schedule control system

CV   cost variance

DE   desired effect

DOD   Department of Defense

F&OR   functional and operational requirements

FMEA   failure modes and effects analysis

FRT   future reality tree

GAO   Government Accounting Office

MORT   management oversight or risk tree

NASA   National Aeronautics and Space Administration

NBR   negative branch

PERT   Program Evaluation and Review Technique

PMBOK™   *Guide to the Project Management Body of Knowledge*

PMP   project management plan

PRT   prerequisite tree

PSA   probabilistic safety assessments

ROI   return on investment

SOW   statement of work

SV   schedule variance

TOC   theory of constraints

TQM   total quality management

TRT   transition tree

UDE   undesirable effect

WBS   work breakdown structure

Glossary

Activity   The lowest level of the work breakdown structure (WBS); a packet of work that forms the basic building block of a plan or a network.

Activity network   A network made up of two or more activities with dependency.

Actual cost   The actual money spent in performing an activity so far.

Actual cost of work performed (ACWP)   The CSCS term for actual cost.

Additional cause reservation   A reservation applied to a thinking process tree suggesting that there may be an additional cause that creates an effect at least one-third of the time.

Banana   Title for the logical "and" on a thinking process tree, represented by a flat oval connecting the causality arrows to be "anded."

Bar chart   See *Gantt chart.*

Body of knowledge   A document produced by the Project Management Institute that defines areas of knowledge that make up the discipline of project management.

Bottleneck   The constraint in a production flow process; the limiting capacity process step; the critical chain in a single-project bottleneck. A company may also have company resource constraints.

Budget   The planned cost for a project; BCWS. It also may include a cost buffer or a budget contingency.

Budgeted cost of work performed (BCWP)   Earned value of work done, equal to the amount that was budgeted for the activities completed.

Budgeted cost of work scheduled (BCWS)   The value of work that should have been completed by the current date according to the baseline plan.

Buffer   Time or budget allowance used to protect scheduled through-put, delivery dates, or cost estimates on a production process or project. Buffers are sized based on the uncertainty in the protected group of activities. Therefore, the schedule buffers are not the same as float or slack that occurs as an accident of the activity logic in critical path schedules.

Buffer penetration   The amount of the buffer that has been used up by actual progress in the project.

C/SCSC   See *cost/schedule control system criteria.*

Capacity constraint buffer (CCB)   Buffer placed between projects or in the drum schedule to sequence projects in a multiproject environment.

Capacity constraint resource   In a multiple project environment, the resource that is most often overloaded.

Categories of legitimate reservation (CLR)   A set of logical tests for trees created by the thinking process.

Causality reservation   Questioning if an effect at the head of a causality arrow in a thinking process tree is an unavoidable consequence of the entity or entities at the tail of the arrow.

Cause   An entity that inevitably leads to a certain result (effect). Causality is determined if the predicted effect is always present when the cause is present and never present when it is not. Causes may be single or may require other conditions to lead to the effect.

Clarity reservation   One of the legitimate reservations for scrutinizing thinking process trees. It means the reviewer does not fully understand the entity.

Cloud (evaporating)   A fixed-format necessity tree used to develop win-win solutions to action alternatives or conflicting wants. The action alternatives are best expressed as opposites, for example, "Do D; do not do D." The cloud has five entities and arrows (see *thinking process*). You identify the assumptions underlying the arrows to resolve the cloud and develop injections that will invalidate the assumption and therefore invalidate the arrow and "dissolve" the cloud.

Common cause   A single entity that causes several effects.

Common cause variation   Variation of process output that is within the capability of the process, and therefore not assignable to a special cause. Also called *natural variation.*

Conflict resolution diagram (CRD)   An alternative title for evaporating cloud.

Constraint   A process or process step that limits throughput.

Core problem   A primary cause of most of the UDE symptoms in your system. You identify the core problem as an entry point on your CRT that traces, in cause-effect-cause relationships, through at least two-thirds of the UDEs, and that you have the stamina and energy to change.

Communication   The effective transmission of information so that the recipient understands clearly what the sender intends. Communication media take several forms: oral, written, textural, numeric, graphic, body language, paper, electronic, physical, and so on. In short, bring together and effectively control those things that need interrelating for the project to be properly assessed, configured, and implemented.

Communication current reality tree (CCRT)   A special form of the current reality tree developed to communicate the tree for buy-in. It displays the cloud for the system at the bottom and builds to the undesired effects.

Communication future reality tree (CFRT)   A special form of the future reality tree developed for buy-in of a specific group of people. It shows the connection between the injections and specific benefits for the group addressed.

Conflict management   The art of managing conflict effectively. In the thinking process, conflict management involves the evaporating cloud, communication transition trees, and, for chronic conflicts, the negative branch.

Constraints   In a project, the generic term for factors that affect the possible start and finish dates of an activity, including logic and imposed dates. In the theory of constraints, the factor that limits the system from obtaining more throughput.

Contingency   The difference between a 90-95% probable estimate and a 50% probable estimate.

Core conflict   The conflict that leads to the core problem.

Core problem   A problem that causes at least two-thirds of the undesired effects in a current reality tree and that you have the stamina and energy to reverse. The core problem is often the root cause of a number of root causes or the common cause.

Corrective action   A process for correcting defects by identifying the defect, assigning responsibility, performing causal analysis, planning a resolution, and implementing the resolution.

Cost buffer   The financial contingency added to a project to protect the overall project cost. As with schedule buffers, it is best to accumulate all the individual activity cost contingencies into one, that will be much smaller than the sum of the individual buffer.

Cost schedule control system (CSCS)   A system for evaluating the work completed on a project as a basis for progress payments. The primary innovation is the use of BCWP, that is, the estimate for an activity, as the measure of work completed.

Cost/schedule control systems criteria (C/SCSC)   In 1967, the U.S. Department of Defense defined a standard for the use of earned value analysis in defense projects. C/SCSC has since been adopted much more widely and is supported by most planning software.

Cost variance   The value of the work done less the actual cost of the work done, that is, BCWP–ACWP. A negative number shows that the project is currently over budget.

Cost world   A business perspective that focuses on reducing cost as the path to business success. The cost world is typified by a belief that cost savings are additive.

CPM   The critical path method, the original innovation in using networks and defining a critical path through the network.

Critical activity   An activity on the critical chain.

Critical chain   The longest set of dependent activities, with explicit consideration of resource availability, to achieve a project goal. The critical chain is *not* the same as what you get from performing resource allocation on a critical path schedule. The critical chain defines an alternate path that completes the project earlier by resolving resource contention upfront.

Critical chain feeding buffer (CCFB)   A time buffer at the end of a project activity chain that feeds the critical chain.

Critical chain project management (CCPM)   A project management system that addresses all the UDEs from the project management CRT. It includes a critical chain plan, the flush measurement tool, buffer management, and roadrunner task performance.

Critical chain resource buffer (CCRB)   See *resource flag.*

Critical chain schedule   A late finish plan controlled by the critical chain, including a project buffer, critical chain feeding buffers, and resource buffers.

Critical path   The longest sequence of activities in a network. Usually, but not always, a sequence with zero float. The critical path is an accident of arithmetic. It may be the longest sequence of activities, but there may be others that have such minimal float as to be inconsequential. It also does not account for resource constraints. Once resource leveling has been performed, slack and the critical path are no longer valid calculations. (All paths ususally contain gaps.) The Project Management

Institute's definition of critical path notes that it changes as the project progresses.

CSCS   Cost schedule control system.

Current reality tree (CRT)   A logical representation of the current business system under analysis, demonstrating how the core conflict connects to the system UDEs.

CV   See *cost variance.*

Data   Any string of characters that describes something about our reality.

Dependency links   The various types of link connecting activities in a precedence network; include finish to start, start to start, finish to finish, and start to finish.

Dependent events   Events or effects that are related in magnitude, time, or some other factor such that they influence each other or have a common cause influence; events in which the output of one event influences the input to another event.

Desired effect (DE)   The positive effect you want to have in future reality to replace your undesired effect of current reality.

Dollar-days   The integration of the product of daily cash flow (in minus out) times the number of days. See *flush.*

Drum   The resource selected for sequencing of projects. See *capacity constraint resource.* In production, it is the bottleneck processing rate, which is used to schedule an entire plant.

Drum buffer   A buffer placed in a project plan to ensure that the company constraint is not starved. Also known as a constraint resource buffer, strategic resource buffer, or constraint buffer.

Drum-buffer-rope   Method for production scheduling. The drum is the capacity of the plant constraint and is used to set the overall throughput schedule. The buffers are in-process inventories strategically located to eliminate starving the constraint due to statistical fluctuations. The rope is the information connection between the constraint and material release into the process.

Duration The amount of elapsed time an activity is estimated to take.

Early finish date The earliest date by which an activity can finish. Calculated during the forward pass of critical path analysis.

Early start date The earliest date by which an activity can start. Calculated during the forward pass of critical path analysis.

Earned value The value of the work done in which value is calculated in terms of the baseline cost. Known as BCWP in the C/SCSC.

Earned value analysis The analysis of project progress in which the actual money spent is compared to the value of the work achieved. See also *cost/schedule control systems criteria.*

Effect An entity representing the result of one or more causes.

Efficiency A measure of the speed and effectiveness with which a resource delivers a particular skill; a measure of how much time resources charge to projects versus unbillable time.

Elevate TOC term for increasing the throughput capability of the system constraint. For projects, that usually means adding resources.

Entity A condition that exists.

Entry point An entity on a sufficiency tree that has no causes (arrows) leading into it.

Erroneous information A wrong answer to the question asked.

Estimate at completion The current estimated total cost of the project.

Estimate to complete (ETC) An estimate of the time or effort required to complete the activity.

Estimating The process of developing the planned cost and duration for activities.

ETC See *estimate to complete.*

Evaporating cloud See *cloud.*

Existence reservation Means, "prove it"; can be applied to an entity or a causality arrow in a thinking process tree.

Exploit   TOC term for ensuring that the system makes most effective use of a constraint in terms of the system goal.

External constraint   Constraints that act on activities in a network from outside the network, typically regulations, imposed dates, or an environmental condition.

Feeding buffer   See *critical chain feeding buffer (CCFB)*.

Finish to start   A type of dependency link in precedence networks that indicates that the start of the successor activity may not occur until the predecessor activity has finished.

Five focusing steps   The five-step process to identify and elevate constraints. (See Chapter 1.)

Float   A measure of the time flexibility available in the performance of an activity. Available in three flavors: total float, free float, and independent float. The minimum amount of time by which an activity will be extended due to factors outside the project manager's control. See *slack*.

Free float   The amount of time an activity may be delayed without causing delay to successor activities.

FRT   Future reality tree.

Future reality tree (FRT)   A sufficiency tree connecting injections to desired effects.

Gantt chart   A chart showing a list of activities represented by bars that are proportional in length to their duration. The bars are positioned along a horizontal time scale.

Goal   See *Jonah*.

Hockey stick   The shape of a curve that is relatively flat and then rises rapidly, representing, for example, the amount of effort one puts out as a deadline approaches.

House-on-fire reservation   Original definition based on the logic statement, "*If* there are smoke and fire engines, *then* the house is on fire." The smoke and fire engines are not the cause of the house being on fire, but rather cause us to know that the house is on fire. Because the thinking process trees are effect-cause-effect trees, house-on-fire is not correct.

Identify   The first step of the TOC focusing process, consisting of identifying the system constraint.

Information   An answer to the question asked.

Injection   An action or effect that will be created in the future to change system performance.

Integrated plan   A plan combining cost and schedule to complete a project.

Intermediate objective (IO)   An action or effect that is a necessary prerequisite to an injection or another IO.

Invalid data   Data that is not needed to deduce the specific desired information.

Inventory   All the investment in the equipment necessary to convert raw material into throughput.

Jonah   A title bestowed on those who complete the AGI Jonah course and are therefore prepared to go forth and replenish the rain forests with trees. A leading character in Dr. Goldratt's book *The Goal*. Jonah is a teacher and leader in the Socratic tradition.

Late-finish date   The latest date by which an activity can finish. Calculated during the backward pass of critical path analysis. All activities in a critical chain schedule use this date, except those moved forward in time to resolve resource contention.

Leadership   Doing the right things and getting others to follow.

Linked projects   Term used in some computer packages to indicate projects that use a common set of resources.

Logic link   See *dependency links*.

Logic loop   A circular sequence of dependency links between activities in a network.

Mean   The average of a group of data, also called the first moment of the data. In a distribution skewed to the right, like most duration and cost estimates, the mean is higher than the median.

Median   The middle value in a group of ordered data.

Merge node   A node in a network diagram in which two or more links or activities merge.

Milestone   An activity of zero duration that represents a significant deliverable or stage of the project.

Milestone plan   A plan that contains only milestones that highlight key points in the project.

Multiproject management   The art and science of managing multiple projects that are, in some way, interconnected. These may be logic connections or, more likely, use of common resources.

Multitasking   Performing more than one project activity at the "same" time.

Necessary condition 1   Satisfy customers now and in the future. (A necessary condition to meet the goal of any enterprise.)

Necessary condition 2   Satisfy and motivate employees now and in the future. (A necessary condition to meet the goal of any enterprise.)

Necessity tree   A logic tree in which each item at the tail of an arrow must exist for the item at the head of the arrow to exist, because of some assumption or obstacle represented by the arrow.

Need   The requirement(s) that must be met to achieve an objective or goal.

Negative branch   A sufficiency logic tree (potential FRT) stemming from an injection; which may lead to UDEs.

Network   A diagram in which the logical relationships between activities are shown in either activity on arrow or precedence format.

Network analysis   Generic term for analyzing networks including PERT and critical path analysis.

Node   The start and end of activities in an activity on an arrow network or the activity box in a precedence network.

Obstacle   An entity that prevents an effect from existing.

Operating expense   All the money it costs to convert raw material into throughput.

Percentage complete   A number estimating the amount of an activity that is finished; one of the ways of allocating BCWP.

Performance measurement   Method used to relate physical progress achieved with cost status; identifies whether cost variances are due to differences in the value of the work being performed being too expensive or under budget. In that way, it is possible to determine if a project is ahead, on, or behind budget. See *earned value analysis.*

PERT   See *program evaluation and review technique.*

Pessimistic duration   The longest of the three durations in the three-duration technique or PERT.

Plan   Generic term used for a statement of intentions whether they relate to time, cost, or quality in their many forms.

PMI   Project Management Institute.

PMP   Project management plan.

Predecessor   An activity that logically precedes the current activity. See also *successor.*

Predicted effect reservation   One of the categories of legitimate reservations; means, "that entity can't be right, because if it existed we would see another predicted effect."

Prerequisite tree (PRT)   A logic tree representing the time phasing of actions to achieve a goal, connecting intermediate objectives with effects that overcome obstacles. The PRT is read, "to have *entity at head of arrow*, we must have *entity at tail of arrow* because of *obstacle*."

Priority   Means of defining the order in which activities will be scheduled during resource scheduling.

Probability   Usually used in the context of risk as a measure of the likelihood of a risk occurring.

Problem   Gap between what we want and what we have.

Process   Sequence of interconnected activities, each of which has an input and an output.

Program   Portfolio of projects selected and planned in a coordinated way so as to achieve a set of defined objectives, giving effect to various (and often overlapping) initiatives and/or implementing a strategy; alternatively, a single large or very complex project or a set of otherwise unrelated projects bound by a business cycle.

Program evaluation and review technique (PERT)   Network scheduling tool, initially distinguished from CPM by allowing and using three activity-duration estimates.

Program management   The selection and coordinated planning of a portfolio of projects so as to achieve a set of defined business objectives and the efficient execution of those projects within a controlled environment such that they realize maximum benefit for the resulting business operations.

Program manager   The individual responsible for day-to-day management of the program.

Program plan   A plan for a program of projects; distinguished from a program management plan in that a program plan need not supply the management systems.

Progress reporting   Process of gathering information on work done and revised estimates, updating the plan, and reporting the revised plan.

Project   A temporary management environment created to achieve a particular business objective through the control and coordination of logistical and technical resources.

Project buffer   Time buffer placed at the end of the critical chain in a project schedule to protect the overall schedule.

Project management   The managerial task of accomplishing a project on time, within budget, and to technical specification. The project manager is the single point of responsibility for achieving that.

Project manager   The person appointed to take day-to-day responsibility for management of the project throughout all its stages.

Quality   According to Dr. Joseph Juran, "fitness for use"; defined in terms of both a lack of defects and product features. Phillip Crosby defined it as "conformance to customer requirements." W. Edwards Deming stated, "A product or service possesses quality if it helps somebody and enjoys a good and sustainable market."

Required data   The data needed by the decision procedure to derive information.

Resource   Entity that performs project work, comprising people, contractors, and machines.

Resource buffer   Flag placed on the critical chain to ensure that resources are available when needed to protect the critical chain schedule. The flag is insurance of resource availability and does not add time to the critical chain. It takes the form of a contract with the resources that ensures their availability, whether or not you are ready to use them then, through the latest time you might need the resource. Often called a critical chain resource buffer (CCRB).

Resource leveling   The process of rescheduling activities such that the requirement for resources on the project does not exceed resource limits.

Resource limit   The amount of a particular resource available to the project at a point in time.

Risk   The combination of probability and consequence of an undesired outcome. Project risk usually denotes undesired outcome relative to the project scope, cost, or schedule. Other risks sometimes important to projects are safety, environment, business, and security risks.

Root cause   The cause that, if changed, will prevent recurrence of an UDE.

Rope   The information flow from the drum (bottleneck or constraint resource) to the front of the line (material release) that controls plant production.

Schedule   Collection of reports showing the timing of activities.

Schedule variance   Value of the work done less the value of the work that should have been done, that is, BCWP – BCWS. A negative number shows that the task is behind schedule. When rolled up for multiple tasks, it shows nothing or erroneous information about project schedule performance.

Scheduling   Applying start and finish dates to project tasks.

Scrutiny   Inspection of a tree to ensure that none of the categories of legitimate reservation applies and that all the entities are necessary to connect the undesired effects.

Slack   Free time in a critical path schedule resulting from paths shorter than the critical path. See *float*.

Special cause variation   Variation in the output of a process that has an assignable cause.

Statistical fluctuations   Common cause variations in output quantity or quality, including activity duration and cost.

Student syndrome   The natural tendency of many people to wait until a due date is near before applying full energy to complete an activity. Also see *hockey stick*.

Sufficiency tree   A tree construction in which the existence of the entities at the tail of the arrow makes the existence of the entity at the head of the arrow an unavoidable result.

Subordinate   The third step in the TOC five-step focusing process, placing considerations not related to the company goal at lower importance than items that directly affect the system's ability to achieve the goal.

Successor   An activity that logically succeeds the current activity. See also *predecessor*.

System   A network of interdependent components that work together to accomplish the aim of the system. Without an aim, there is no system.

Task   A term usually synonymous with *activity*.

Theory of Constraints (TOC)   A system theory developed by Dr. Eliyahu Goldratt and first published in his book *The Goal*. The most basic statement of the theory is that the output of a system is limited by a constraint.

Thinking process   The five-step process that identifies what to change, what to change to, and how to cause the change. The sequence of steps starts with the CRT; goes through the evaporating cloud, FRT, and PRT; and results in the TRT.

Throughput   All the money customers pay minus the raw material cost.

Transition tree (TRT)   Plan that specifies the effects to be achieved, the starting conditions, the actions necessary to create the effects, the logic of why the action will create the effect, and the logic for the sequence of the effects.

Variance (statistical)   A measure of the dispersion of a sample and an estimate of the standard deviation of a population.

Want   The effect that one believes must exist to satisfy a need, because of some set of assumptions.

WBS   Work breakdown structure.

Work breakdown structure (WBS)   A tree diagram that breaks a project down into increasing levels of detail. The lowest level of the work breakdown structure comprises work packages.

About the author

Larry Leach is the principal of Quality Systems, a management consulting firm. Quality Systems focuses on applying successful and logical business tools to help clients improve their work processes and management systems. Quality Systems specializes in leading the implementation of the new Critical Chain method of project management. Prior to founding Quality Systems, Larry worked at the vice presidential level in several Fortune 500 companies, and was a systems analysis division director for the U. S. Department of Energy. Larry's 25 years of experience as a project manager include projects ranging from research and development to construction.

Larry Leach has master's degrees in both business management and mechanical engineering from the University of Idaho and the University of Connecticut, respectively. He was awarded a membership in Tau Beta Pi, the Engineering Honorary Society while earning his undergraduate degree and the Stevens Institute of Technology. Mr. Leach is a member of the Project Management Institute and the American Society for Quality Control. He has published many papers on related topics in his field, and in addition to publishing the *Critical Chain Project Management* (Artech House, 2000), he is the self-published author of the *The Critical Chain Project Manager's Fieldbook*.

Index

A

B

For further information on these and other Artech House titles, including previously considered out-of-print books now available through our In-Print-Forever® (IPF®) program, contact:

Artech House
685 Canton Street
Norwood, MA 02062
Phone: 781-769-9750
Fax: 781-769-6334
e-mail: artech@artechhouse.com

Artech House
46 Gillingham Street
London SW1V 1AH UK
Phone: +44 (0)20 7596-8750
Fax: +44 (0)20 7630-0166
e-mail: artech-uk@artechhouse.com

Find us on the World Wide Web at:
www.artechhouse.com